Math Word Problems

DeMYSTiFieD®

DeMYSTiFieD® Series

Math Word Problems
DeMYSTiFieD®

Allan G. Bluman

Second Edition

New York   Chicago   San Francisco   Lisbon   London   Madrid   Mexico City
Milan   New Delhi   San Juan   Seoul   Singapore   Sydney   Toronto

The McGraw·Hill Companies

Cataloging-in-Publication Data is on file with the Library of Congress.

McGraw-Hill books are available at special quantity discounts to use as premiums and sales promotions, or for use in corporate training programs. To contact a representative please e-mail us at bulksales@mcgraw-hill.com.

Math Word Problems DeMYSTiFieD®, Second Edition

1 2 3 4 5 6 7 8 9 0   DOC/DOC   1 9 8 7 6 5 4 3 2 1

ISBN 978-0-07-176386-8
MHID 0-07-176386-4

Sponsoring Editor Judy Bass	**Production Supervisor** Pamela A. Pelton
Editing Supervisor David E. Fogarty	**Composition** Cenveo Publisher Services
Project Manager Sandhya Gola, Cenveo Publisher Services	**Art Director, Cover** Jeff Weeks
Copy Editor Lunaea Weatherstone	**Cover Illustration** Lance Lekander
Proofreader Sheena Uprety, Cenveo Publisher Services	

To Betty Claire, Allan, Mark, and all my students
who have made my teaching career an enjoyable experience

About the Author

Allan G. Bluman taught mathematics and statistics in high school, college, and graduate school for 39 years. He received his doctorate from the University of Pittsburgh. He has written three mathematics textbooks published by McGraw-Hill. He is also the author of three other mathematics books in the McGraw-Hill DeMYSTiFieD series: *Pre-Algebra DeMYSTiFieD*, *Probability DeMYSTiFieD*, and *Business Math DeMYSTiFieD*. He is the recipient of "An Apple for the Teacher" award for bringing excellence to the learning environment and two "Most Successful Revision of a Textbook" awards from McGraw-Hill. His biographical record appears in *Who's Who in American Education*, 5th edition. He has been inducted into the McKeesport High School Alumni Hall of Fame.

Contents

Introduction

What did one mathematics book say to another one?
"Boy, do we have problems!"

All mathematics books have problems, and most of them have word problems. Many students have difficulties when attempting to solve word problems. One reason is that they do not have a specific plan of action. A mathematician, George Polya (1887–1985), wrote a book entitled *How to Solve It*, explaining a four-step process that can be used to solve word problems. This process is explained in Chapter 1 of this book and is used throughout the book. This process provides a plan of action that can be used to solve word problems found in all mathematics courses.

This book is divided into 12 chapters. Chapters 1, 2, 3, and 4 explain how to use the four-step process to solve word problems in arithmetic or pre-algebra. Chapter 5 reviews equations and explains algebraic representation. Chapters 6 through 11 explain how to use the process to solve problems in algebra, and these chapters cover all of the basic types of problems (coin, mixture, finance, etc.) found in an algebra course. Chapter 12 explains how to solve word problems in geometry, probability, and statistics. This book also contains six "Refreshers." These are intended to provide a review of topics needed to solve the word problems that follow them. They are not intended to teach the topics from scratch. You should refer to appropriate textbooks if you need additional help with the refresher topics.

This book can be used either as a self-study book or as a supplement to your textbook. You can select the chapters that are appropriate for your needs.

Curriculum Guide

The *DeMYSTiFied*® books are closely linked to the standard high school and college curricula, so the Curriculum Guide on the inside back cover is provided for you to have a clear path to meet your mathematical goals. What many students do not know is that mathematics is a *hierarchical* subject. What this means is that before you can be successful in algebra, you need to know basic arithmetic, since the concepts of arithmetic (pre-algebra) are used in algebra. Before you can be successful in trigonometry, you need to have a basic understanding of algebra and geometry, since trigonometry uses concepts from these two courses. You can use this Guide in your mathematical studies to learn which courses are necessary before taking the next ones.

How to Use This Book

As you know, in order to build a tall building, you need to start with a strong foundation. The same is true when mastering mathematics. This book presents the basic types of mathematical word problems and how to solve them in a logical, easy-to-read format. This book can be used as an independent study course or as a supplement to other mathematical courses.

To learn how to solve word problems, you must know the basic procedures and be able to apply these procedures to mathematical word problems. This book is written in a style that will help you with learning. As stated previously, it follows the basic problem-solving strategy stated by George Polya. It also contains six mathematical refreshers to help you review topics that are used in word-problem solving. Basic facts and helpful suggestions can be found in the "Still Struggling" boxes. Each section has several worked-out examples showing you how to use the rules and procedures. Each section also contains several practice problems for you to work out to see if you understand the concepts. The correct answers are provided immediately after the problems so that you can see if you have solved them correctly. At the end of each chapter, there is a multiple-choice quiz. If you answer most of the problems correctly, you can move on to the next chapter. If not, you can repeat the chapter. Make sure that you do not look at the answer before you have attempted to solve the problem.

Even if you know some or all of the material in the chapter, it is best to work through the chapter in order to review the material. The little extra effort will be a great help when you encounter the more difficult material later. After you complete the entire book, you can take the 50-question final exam and deter-mine your level of competence. It is suggested that you use a calculator to help you with the computations.

I would like to answer the age-old question, "Why do I have to learn this stuff?" There are several reasons. First, mathematics is used in many academic fields. If you cannot do mathematics, you severely limit your choices of an academic major. Second, you may be required to take a standardized test for a job, degree, or graduate school. Most of these tests have a mathematical section. Third, a working knowledge of word problems will go a long way to help you solve mathematical problems that you encounter in everyday life. I hope this book will help you learn mathematics.

For the second edition, most of the examples and exercises have been changed. Also, at the beginning of each chapter, the basic objectives have been stated and a brief summary appears at the end of the chapter. In addition, the "Still Struggling" explanation boxes have been added. The section on mixture problems has been rewritten to explain the ideas more clearly. In the section on probability problems, the sample space for cards has been added, and the four basic rules for probability have been included.

Best wishes on your success.

Allan G. Bluman

Acknowledgments

I would like to thank my wife, Betty Claire, for helping me with this project, and I wish to express my gratitude to my editor, Judy Bass, and to Carrie Green for her suggestions and error checking.

Math Word Problems

DeMYSTiFieD®

chapter **1**

Introduction to Problem Solving

This chapter explains the basic four-step problem-solving technique developed by George Polya. In addition, some basic problem-solving strategies such as drawing a picture, making a list, etc., are explained.

CHAPTER OBJECTIVES

In this chapter, you will learn how to

- Use the four-step problem-solving method
- Solve word problems using general problem-solving strategies

Four-Step Method

In every area of mathematics, you will encounter "word" problems. Some students are very good at solving word problems while others are not. When teaching word problems in pre-algebra and algebra, I often hear, "I don't know where to begin" or "I have never been able to solve word problems." A great deal has been written about solving word problems. A Hungarian mathematician, George Polya, did much in the area of problem solving. His book, entitled *How to Solve It*, has been translated into at least 17 languages, and it explains the basic steps of problem solving. These steps are explained next.

Step 1: Understand the problem   First read the problem carefully several times. Underline or write down any information given in the problem. Next, decide what you are being asked to find. This will be called the *goal*.

Step 2: Select a strategy to solve the problem   There are many ways to solve word problems. You may be able to use one of the basic operations such as addition, subtraction, multiplication, or division. You may be able to use an equation or formula. You may even be able to solve a given problem by trial and error. This step will be called *strategy*.

Step 3: Carry out the strategy   Perform the operation, solve the equation, etc., and get the solution. If one strategy doesn't work, try a different one. This step will be called *implementation*.

Step 4: Evaluate the answer   This means to check your answer if possible. Another way to evaluate your answer is to see if it is reasonable. Finally, you can use *estimation* as a way to check your answer. This step will be called *evaluation*.

When you think about the four steps, they apply to many situations that you may encounter in life. For example, suppose that you play basketball. The *goal* is to get the basketball into the hoop. The *strategy* is to select a way to make a basket. You can use any one of several methods such as a jump shot, a layup, a one-handed push shot, or a slam dunk. The strategy you use will depend on the situation. After you decide on the type of shot to try, you *implement* the shot. Finally, you *evaluate* the action. Did you make the basket? Good for you! Did you miss it? What went wrong? Can you improve on the next shot?

Now let's see how this procedure applies to a mathematical problem.

EXAMPLE

Find the next two numbers in the sequence

$$5 \quad 12 \quad 8 \quad 15 \quad 11 \quad 18 \quad 14 \quad \underline{\quad} \quad \underline{\quad}$$

SOLUTION

Goal: You are asked to find the next two numbers in the sequence.

Strategy: Here you can use a strategy called "find a pattern." Ask yourself, "What's being done to one number to get the next number in the sequence?" In this case, to get from 5 to 12, you can add 7. But to get from 12 to 8, you need to subtract 4. So perhaps it is necessary to do two different things.

Implementation: Add 7 to 14 to get 21. Subtract 4 from 21 to get 17. Hence, the next two numbers should be 21 and 17.

Evaluation: In order to check the answers, you need to see if the "add 7, subtract 4" solution works for all the numbers in the sequence, so start with 5.

$$5 + 7 = 12$$
$$12 - 4 = 8$$
$$8 + 7 = 15$$
$$15 - 4 = 11$$
$$11 + 7 = 18$$
$$18 - 4 = 14$$
$$14 + 7 = 21$$
$$21 - 4 = 17$$

Voilà! You have found the solution!
Now let's try another one.

EXAMPLE

Find the next two numbers in the sequence

$$1 \quad 3 \quad 7 \quad 13 \quad 21 \quad 31 \quad 43 \quad \underline{\quad} \quad \underline{\quad}$$

SOLUTION

Goal: You are asked to find the next two numbers in the sequence.

Strategy: Again we will use "find a pattern." Ask yourself, "What is being done to the first number to get the second one?" Here we are adding 2. Does adding 2 to the second number 3 give us the third number 7? No. You must add 4 to the second number to get the third number 7. How do we get from the third number to the fourth number? Add 6. Let's apply the strategy.

Implementation:

$$1 + 2 = 3$$
$$3 + 4 = 7$$
$$7 + 6 = 13$$
$$13 + 8 = 21$$
$$21 + 10 = 31$$
$$31 + 12 = 43$$
$$43 + 14 = 57$$
$$57 + 16 = 73$$

Hence, the next two numbers in the sequence are 57 and 73.

Evaluation: Since the pattern works for the first seven numbers in the sequence, we can extend it to the next two numbers, which then makes the answers correct.

EXAMPLE

Find the next two letters in the sequence

A   Z   C   Y   E   X   G   W   _____   _____

SOLUTION

Goal: You are asked to find the next two letters in the sequence.

Strategy: Again, you can use the "find a pattern" strategy. Notice that the sequence starts with the first letter of the alphabet, A, and then goes to the last letter, Z, then back to C, and so on. So it looks like there are two sequences.

Implementation: The first sequence is A C E G, and the second sequence is Z Y X W. Hence, the next two letters are I and V.

Evaluation: Putting the two sequences together, you get A Z C Y E X G W I V.

Now you can try a few problems to see if you understand the problem-solving procedure. Be sure to use all four steps.

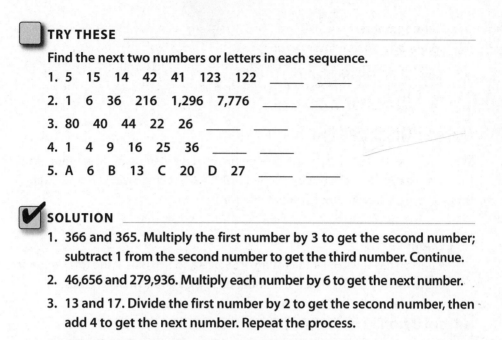

TRY THESE

Find the next two numbers or letters in each sequence.

1. 5   15   14   42   41   123   122   _____   _____

2. 1   6   36   216   1,296   7,776   _____   _____

3. 80   40   44   22   26   _____   _____

4. 1   4   9   16   25   36   _____   _____

5. A   6   B   13   C   20   D   27   _____   _____

SOLUTION

1. 366 and 365. Multiply the first number by 3 to get the second number; subtract 1 from the second number to get the third number. Continue.

2. 46,656 and 279,936. Multiply each number by 6 to get the next number.

3. 13 and 17. Divide the first number by 2 to get the second number, then add 4 to get the next number. Repeat the process.

4. 49 and 64. Square the numbers in the sequence: 1, 2, 3, 4, …

5. E and 34. Use the alphabet and add 7 to each number.

Well, how did you do? You have just had an introduction to systematic problem solving. The remainder of this book is divided into three parts. Chapters 2–5 explain how to solve word problems in arithmetic and pre-algebra. Chapters 6–11 explain how to solve word problems in introductory and intermediate algebra. Chapter 12 explains how to solve word problems in geometry, probability, and statistics. After successfully completing this book, you will be well along the way to becoming a competent mathematical word problem solver.

Problem-Solving Strategies

There are some general problem-solving strategies you can use to solve real-world problems and help you check your answers when you use the strategies presented later in this book. These strategies can help you with problems found on standardized tests, in other subjects, and in everyday life.

These strategies are

1. Make an organized list
2. Guess and test

3. Draw a picture

4. Find a pattern

5. Solve a simpler problem

6. Work backwards

Make an Organized List

When you use this strategy, you make an organized list of possible solutions and then systematically work out each one until the correct answer is found. Sometimes it helps to make the list in a table format.

EXAMPLE

A person has seven bills consisting of $5 bills and $10 bills. If the total amount of the money is $50, find the number of $5 bills and $10 bills he has.

 SOLUTION

Goal: You are being asked to find the number of $5 bills and $10 bills the person has.

Strategy: This problem can be solved by making an organized list and finding the total amount of money you have as shown:

$5 bills	$10 bills	Amount
1	6	$65

One $5 bill and six $10 bills make seven bills with a value of 1 × $5 + 6 × $10 = $65. This is incorrect, so try two $5 bills and five $10 bills and keep going until a sum of $50 is reached.

Implementation: Finish the list.

$5 bills	$10 bills	Amount
1	6	$65
2	5	$60
3	4	$55
4	3	$50

Hence four $5 bills and three $10 bills are needed to get $50.

Evaluation: Four $5 bills and three $10 bills make seven bills whose total value is $50.

EXAMPLE

In a barnyard there are eight animals, chickens and cows. Chickens have two legs and cows have four legs, of course. If the total number of legs is 22, how many chickens and cows are there?

SOLUTION

Goal: You are being asked to find how many chickens and how many cows are in the barnyard.

Strategy: You can make an organized list, as shown.

Chickens (2 legs)	Cows (4 legs)	Total number of legs
1	7	30

The number of chickens and cows must sum to 8 and that gives a total of 30 legs:

$$1 \times 2 + 7 \times 4 = 2 + 28 = 30$$

Implementation: Continue the table until the correct answer (22 legs) is found.

Chickens (2 legs)	Cows (4 legs)	Total number of legs
1	7	30
2	6	28
3	5	26
4	4	24
5	3	22

Hence, there are five chickens and three cows in the barnyard.

Evaluation: Five chickens have $5 \times 2 = 10$ legs, and three cows have $3 \times 4 = 12$ legs, $10 + 12 = 22$ legs.

Guess and Test

This strategy is similar to the previous one except you do not need to make a list. You simply take an educated guess at the solution and then try it out to see if it is correct. If not, try another guess; then test it.

EXAMPLE

The sum of the digits of a two-digit number is 9. If the digits are reversed, the new number is nine more than the original number.

SOLUTION

Goal: You are being asked to find a two-digit number.

Strategy: You can use the guess and test strategy. First guess some two-digit numbers such that the sum of the digits is 9. For example, 18, 27, 36, 45, etc., meet this part of the solution. Then see if they meet the other condition of the problem.

Implementation:

Guess: 27; reverse the digits: 72; subtract: $72 - 27 = 45$

Guess: 36; reverse the digits: 63; subtract: $63 - 36 = 27$

Guess: 45; reverse the digits: 54; subtract: $54 - 45 = 9$. This is the correct solution; hence, the number is 45.

Evaluation: The sum of the digits $4 + 5 = 9$, and the difference $54 - 45 = 9$.

EXAMPLE

The letters *X* and *W* each represent a digit from 0 through 9. Find the value of each letter so that the following is true:

$$
\begin{array}{r}
X \\
X \\
+X \\
\hline
WX
\end{array}
$$

SOLUTION

Goal: You are being asked to find what digits *X* and *W* represent.

Strategy: Use guess and test.

Implementation: Guess a few digits for *X* and see what works:

$$
\begin{array}{ccc}
X = 4\text{: } 4 & X = 7\text{: } 7 & X = 5\text{: } 5 \\
4 & 7 & 5 \\
+4 & +7 & +5 \\
\hline
12 & 21 & 15
\end{array}
$$

Hence $X = 5$ and $W = 1$ is the correct answer.

Evaluation: Notice that all the digits in the column are the same; that is, they are all the same number. You must add three single-digit numbers and get the same number as the one's digit of the solution. There are only two possibilities: 0 and 5. Since the answer has two digits, 0 is disregarded.

Draw a Picture

Many times a problem can be solved using a picture, figure, or diagram. Also, drawing a picture can help you to determine which other strategy can be used to solve a problem.

 EXAMPLE

Ten trees are planted in a row at three-foot intervals. How far is it from the first tree to the last tree?

SOLUTION

Goal: You are being asked to find the distance from the first tree to the last tree.

Strategy: Draw a figure and count the intervals between them; then multiply the answer by 3.

Implementation: Solve the problem. See Figure 1-1.

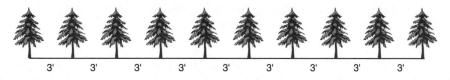

FIGURE 1-1

Since there are nine intervals, the distance between the first and last one is 9 × 3 = 27 feet.

Evaluation: The figure shows that 27 feet is the correct answer.

 EXAMPLE

A family has three children. List the number of ways according to gender that the births can occur.

SOLUTION

Goal: You are being asked to list the total number of ways three children can be born.

Strategy: Draw a diagram showing the way the children can be born.

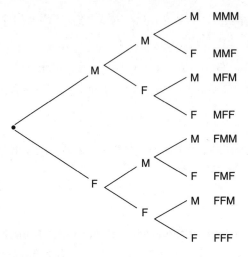

FIGURE 1-2

Implementation: Each child could be born as a male or a female. See Figure 1-2. Hence there are eight different possibilities:

MMM	FMM
MMF	FMF
MFM	FFM
MFF	FFF

Evaluation: Since there are two ways for each child to be born, there are $2 \times 2 \times 2 = 8$ different ways that the births can occur.

Find a Pattern

Many problems can be solved by recognizing that there is a pattern to the solution. Once the pattern is recognized, the solution can be obtained by generalizing from the pattern.

EXAMPLE

A wealthy person decided to pay an employee $1 for the first day's work, $2 for the second day's work, and $4 for the third day's work, etc. How much did the employee earn for 15 days of work?

SOLUTION

Goal: You are being asked to find the amount the employee earned for a total of 15 days of work.

Strategy: You can make a table starting with the first day and continuing until you see a pattern.

Implementation:

Day	Amount earned	Total amount earned
1	1	1
2	2	3
3	4	7
4	8	15
5	16	31
6	32	63

Notice that the amount earned each day is given by 2^{n-1} where n is the number of the day. For example, on the 6th day, the person earns $2^{6-1} = 2^5 = \$32$. So on the 15th day, a person earns 2^{15-1} or $2^{14} = \$16,384$. The total amount the person earns is given by doubling the amount earned that day and subtracting one. So the total amount earned at the end of the 15 days is $\$16,384 \times 2 - 1 = \$32,767$.

Evaluation: You could check your answer by continuing the pattern for 15 days.

EXAMPLE

Find the answer to $12345678 \times 9 + 9$ using a pattern.

$$1 \times 9 + 2 = 11$$
$$12 \times 9 + 3 = 111$$
$$123 \times 9 + 4 = 1111$$

SOLUTION

Goal: You are being asked to find the answer to $12345678 \times 9 + 9$ using a pattern.

Strategy: Make a table starting with $1 \times 9 + 2$, $12 \times 9 + 3$, $123 \times 9 + 4$, etc. Find the answers to these problems and see if you can find a pattern.

Implementation:

$$1 \times 9 + 2 = 11$$
$$12 \times 9 + 3 = 111$$
$$123 \times 9 + 4 = 1111$$

The pattern shows that you get an answer that has the same number of 1s as the last digit that is added. So the answer to the problem would be a number which has 9 1s, that is, 111,111,111.

Evaluation: Perform the operations on a calculator and see if the answer is correct.

Solve a Simpler Problem

To use this strategy, you should simplify the problem or make up a shorter, similar problem and figure out how to solve it. Then use the same strategy to solve the given problem.

EXAMPLE

If there are 10 people at a tennis court and each person plays a singles tennis match with another person, how many different matches can occur?

SOLUTION

Goal: You are being asked to find the total number of different matches played if everybody plays everybody else one time.

Strategy: Simplify the problem using 4 people, and then try to solve it with 10 people.

Implementation: Assume the 4 people are A, B, C, and D. Then write the different games that would occur.

AB, AC, AD, BC, BD, CD

Hence, with 4 people, there would be 6 different games.

Now call the 10 people A, B, C, D, E, F, G, H, I, and J.

AB	AC	AD	AE	AF	AG	AH	AI	AJ
BC	BD	BE	BF	BG	BH	BI	BJ	
CD	CE	CF	CG	CH	CI	CJ		
DE	DF	DG	DH	DI	DJ			
EF	EG	EH	EI	EJ				
FG	FH	FI	FJ					
GH	GI	GJ						
HI	HJ							
IJ								

There would be 45 different games.

Evaluation: You can solve the problem using a different strategy and see if you get the same answer.

Work Backwards

Some problems can be solved by starting at the end and working backwards to the beginning.

EXAMPLE

Tina went shopping and spent $3 for parking and one-half of the remainder of her money in a department store. Then she spent $5 for lunch. Arriving back home, she found that she had $2 left. How much money did she start with?

SOLUTION

Goal: You are being asked to find how much money Tina started with.

Strategy: Work backwards.

Implementation: Work forward first and then work backwards.

1. Spent $3 on parking. Subtract $3.

2. Spent $\frac{1}{2}$ of the remainder in the department store. Divide by 2.

3. Spent $5 on lunch. Subtract $5.

4. Has $2 left.

Reversing the process:

4. $2

3. Add $5 $2 + $5 = $7

2. Multiply by 2 $7 × 2 = $14

1. Add $3 $14 + $3 = $17

Hence, she started out with $17.

Evaluation: Work the problem forward starting with $17 and see if you end up with $2.

Many times there is no single best strategy to solve a problem. You should remember that problems can be solved using different methods or a combination of methods.

TRY THESE

Use one or more of the strategies shown in the lesson to solve each problem.

1. How many cuts are needed to cut a log into eight pieces?

2. Each letter stands for a digit. All identical letters represent the same digit. Find the solution.

$$AB$$
$$+\,B$$
$$\overline{BA}$$

3. The sum of the digits of a two-digit number is 8. If 36 is subtracted from the number, the answer will be the original number with the digits reversed.

4. A person purchased seven candy bars that cost two different prices, $0.89 and $0.99. How many of each kind did the person purchase if the total cost is $6.43?

5. An 20-inch piece of pipe is cut into two pieces such that one piece is three times as long as the other. Find the length of each piece.

6. How many ways can a committee of four people be selected from six people?

7. Frank wants to shape up for basketball. He decides to cut back by eating two fewer cookies each day for five days. During the five days, he ate a total of 40 cookies. How many did he eat on the first day?

8. A mother is four times as old as her daughter. In 16 years, she will be twice as old as her daughter. Find their present ages.

9. How many ways can four different books be lined up in a row on a shelf?

10. Find the tallest person if Betty is taller than Jan, Sue is shorter than Betty, and Jan is taller than Sue.

SOLUTIONS

1. Strategy: Draw a picture: Seven cuts are needed. See Figure 1-3.

2. Strategy: Guess and test: 89 + 9 = 98

3. Strategy: Guess and test: 62 − 36 = 26

4. Strategy: Make an organized list: 5 candy bars at $0.89 and 2 at $0.99.

5. Strategy: Guess and test: 5 inches and 15 inches

6. Strategy: Make an organized list: 15 ways

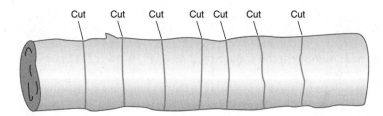

FIGURE 1-3

7. **Strategy: Guess and test: 12 cookies**

8. **Strategy: Make an organized list or guess and test: Mother's age is 32; daughter's age is 8.**

9. **Strategy: Solve a simpler problem: 24 ways**

10. **Strategy: Draw a picture: Betty**

Summary

In this chapter, you have learned the basic word problem-solving procedure that was developed by George Polya. This four-step procedure will be used throughout this book.

Also, you learned some other ways to solve word problems. These ways include making an organized list, guess and test, drawing a picture, finding a pattern, solving a simpler problem using the same strategy on a more difficult problem, and working backwards. There are other strategies that can also be used. They can be found in books on problem solving.

QUIZ

1. The next number in the sequence 3 8 6 11 9 14 12 17 is
   A. 22
   B. 15
   C. 21
   D. 14

2. The next number in the sequence 1 2 3 4 6 7 10 11 15 16 is
   A. 17
   B. 18
   C. 20
   D. 21

3. The next number in the sequence 3 6 4 8 6 12 10 is
   A. 8
   B. 12
   C. 15
   D. 20

4. The largest number that will divide evenly into 180 and 600 is
   A. 12
   B. 20
   C. 30
   D. 60

5. Mary went shopping and bought some pencils ($1 each), notebooks ($2 each), and pens ($3 each). If she spent a total of $12, how many of each item did she purchase?
   A. 3 pencils, 2 notebooks, and 2 pens
   B. 2 pencils, 3 notebooks, and 1 pen
   C. 4 pencils, 1 notebook, and 2 pens
   D. 1 pencil, 3 notebooks, and 2 pens

6. Four horses ran a race. The brown horse finished ahead of the gray horse but behind the black horse. The white horse finished behind the brown horse but ahead of the gray horse. The white horse finished exactly one horse ahead of the gray horse. What was the finishing order of the horses?
   A. black, brown, white, gray
   B. gray, white, brown, black
   C. brown, white, gray, black
   D. black, white, gray, brown

7. Four students are tossing a baseball to each other. If the ball is tossed between each of the other players one time, how many tosses were made?

    A. 5
    B. 6
    C. 12
    D. 30

8. For a party, a person sets up six card tables and pushes them together in a row. How many people can be seated at the arrangement? (Note: Only one person can sit on each side of a card table.)

    A. 6
    B. 12
    C. 14
    D. 24

9. All whole numbers have factors. The factors of 10 are 1, 2, 5, and 10. These are numbers that divide evenly into 10. The numbers 1, 2, and 5 are called proper factors of 10. The number 6 is called a perfect number since its proper factors add up to 6. (1 + 2 + 3 = 6). What is the next perfect number?

    A. 8
    B. 12
    C. 24
    D. 28

10. A rubber ball bounces up half the previous height it fell. If a rubber ball is dropped from a height of 20 feet, how far did it travel by the time it hits the ground three times?

    A. 35 feet
    B. 30 feet
    C. 50 feet
    D. 40 feet

chapter **2**

Solving Decimal and Fraction Problems

This chapter explains how to determine which operation (addition, subtraction, multiplication, or division) you can use to solve problems in arithmetic or pre-algebra. Also, operations with decimals and fractions are reviewed in two refreshers. Finally, word problems using decimals and fractions are explained.

CHAPTER OBJECTIVES

In this chapter, you will learn how to

- Solve word problems using whole numbers
- Use the rules for adding, subtracting, multiplying, and dividing decimals
- Solve word problems using decimals
- Add, subtract, multiply, and divide fractions and mixed numbers, change fractions to decimals, and change decimals to fractions
- Solve word problems using fractions

Operations

Most word problems in arithmetic and pre-algebra can be solved by using one or more of the basic *operations*. The basic operations are addition, subtraction, multiplication, and division. Sometimes students have a problem deciding which operation to use. The correct operation can be determined by the words in the problem.

Use *addition* when you are being asked to find

the total,

the sum,

how many in all,

how many altogether,

etc.,

and when all the items in the problem are the same type or have the same units.

EXAMPLE

For the years 2000–2009, the number of space launches for each country is United States, 201; Russia, 237; China, 49; Japan, 17; and other countries, 79. Find the total number of space launches for the 10-year period.

SOLUTION

Goal: You are being asked to find the total number of space launches that were conducted from 2000 to 2009.

Strategy: Use addition since you need to find a total and all the items in the problem are the same (i.e., space launches).

Implementation: $201 + 237 + 49 + 17 + 79 = 583$.

Evaluation: The total number of space launches is 583. This can be checked by *estimation*. Round each value and then find the sum: $200 + 240 + 50 + 20 + 80 = 590$. Since the estimated sum is close to the actual sum, you can conclude that the answer is probably correct. (Note: When using estimation, you cannot be 100 percent sure your answer is correct since you have used rounded numbers.)

Use *subtraction* when you are asked to find

how much more,

how much less,

how much larger,

how much smaller,

how many more,

how many fewer,

the difference,

the balance,

how much is left,

how far above,

how far below,

how much further,

etc.,

and when all the items in the problem are the same or have the same units.

EXAMPLE

If the highest temperature recorded in Africa was 136°F, and the highest temperature recorded in South America was 120°F, how much higher was the highest temperature in Africa compared to South America?

SOLUTION

Goal: **You are being asked to find how much higher is the highest temperature in Africa compared to the highest temperature in South America.**

Strategy: **Since you are being asked "how much higher" and both items are the same (degrees), you use subtraction.**

Implementation: **136°F − 120°F = 16°. Hence the highest temperature in Africa was 16° higher than the highest temperature recorded in South America.**

Evaluation: **You can check the solution by adding: 120° + 16° = 136°.**

Use *multiplication* when you are being asked to find

the product,

the total,

how many in all,

how many altogether,

etc.,

and when you have groups of individual items.

EXAMPLE

Find the total cost of 15 digital cameras if each one costs $159.

SOLUTION

Goal: You are being asked to find the *total* cost of 15 digital cameras.

Strategy: Use multiplication since you are asked to find a total and you have 15 cameras costing $159 each.

Implementation: $159 × 15 = $2,385. Hence, the total cost of 15 digital cameras is $2,385.

Evaluation: You can check your answer by estimation: 160 × 15 = $2,400. Since $2,400 is close to $2,385, your answer is probably correct.

Use *division* when you are given the total number of items and a number of groups and need to find how many items in each group, or when you are given the total number of items and the number of items in each group and need to find how many groups there are.

EXAMPLE

The shipping department of a business needs to ship 192 pairs of children's shoes. If they are packed 12 pairs per box, how many boxes will be needed?

SOLUTION

Goal: You are being asked to find how many boxes are needed.

Strategy: Here you are given the *total* number of pairs of shoes, 192, and the company needs to pack 12 pairs in each box. You are asked to find how many boxes (groups) are needed. In this case, use division.

Implementation: 192 ÷ 12 = 16 boxes. Hence, you will need 16 boxes.

Evaluation: Check: 16 boxes × 12 pairs of shoes per box = 192 pairs of shoes.

Now you can see how to decide what operation to use to solve arithmetic or pre-aglebra problems.

TRY THESE

1. If seven mountain bicycles cost $1,288, how much does each one cost?

2. If you can burn 12 calories by running at a brisk pace for 1 minute, how many calories can you burn if you run for 20 minutes?

3. A salesperson travels the following miles during a four-day trip:

Monday	852 miles
Tuesday	347 miles
Wednesday	521 miles
Thursday	276 miles

Find the total number of miles the salesperson traveled on the trip.

4. For a specific year, Facebook had 92,208,000 visitors. The MySpace website had 27,966,000 fewer visits during that year. How many visitors did MySpace have?

5. If the average yearly phone bill for a specific year is $588, what is the monthly rate for the phone service?

6. A book company ships its books in boxes that hold 24 books. How many boxes are needed to ship 336 books?

7. Bill purchases eight video games for $18 each. Find the total amount he spent for the games.

8. If you had $357 in your checking account, and you wrote checks for $81 and $116, what would your balance be?

9. Adam decides to save $130 each month for a year. How much money will he have at year's end?

10. A business person mailed five packages costing $8, $14, $18, $3, and $6. Find the total cost of the postage bill.

SOLUTIONS

1. $1,288 ÷ 7 = $184

2. 12 × 20 = 240 calories

3. 852 + 347 + 521 + 276 = 1,996 miles

4. 92,208,000 − 27,966,000 = 64,242,000

5. $588 ÷ 12 = $49

6. 336 ÷ 24 = 14 boxes

7. $18 × 8 = $144

8. $357 − $81 − $116 = $160

9. $130 × 12 = $1,560

10. $8 + $14 + $18 + $3 + $6 = $49

? Still Struggling

If you get the wrong answer, there are two places you could have made a mistake. First, you could have performed the wrong operation. That is, maybe you divided when you should have multiplied. Second, you could have made a mistake in performing the operation or perhaps pressing the wrong key if you are using a calculator. It is best to do the problem over rather than trying to find your mistake. This method works best if the problem requires several steps as the ones found in later chapters in the book.

Refresher I: Decimals

To add or subtract decimals, place the numbers in a vertical column and line up the decimal points. Add or subtract as usual and place the decimal point in the answer directly below the decimal points in the problem.

EXAMPLE

Find the sum: 98.145 + 6.8372 + 421.6

SOLUTION

$$
\begin{array}{r}
98.1450 \\
6.8372 \\
+\ \underline{421.6000} \\
526.5822
\end{array}
$$

Zeros can be written to keep the columns in line

EXAMPLE

Subtract 351.2 − 45.18

SOLUTION

$$
\begin{array}{r}
351.20 \\
-\ \underline{45.18} \\
306.02
\end{array}
$$

To multiply two decimals, multiply the numbers as is usually done. Count the number of digits to the right of the decimal points in the problem and then have the same number of digits to the right of the decimal point in the answer.

■ **EXAMPLE**

Multiply 53.61 × 4.8

✔ **SOLUTION**

$$
\begin{array}{r}
53.61 \\
\times\ 4.8 \\
\hline
42888 \\
21444\ \ \ \\
\hline
257.328
\end{array}
$$

You need
three decimal
places in the
answer

To divide two decimals when there is no decimal point in the divisor (the number outside the division box), place the decimal point in the answer directly above the decimal point in the dividend (the number under the division box). Divide as usual.

■ **EXAMPLE**

Divide 2511.2 ÷ 43

✔ **SOLUTION**

$$
\begin{array}{r}
58.4 \\
43\,)\overline{2511.2} \\
\underline{215}\ \ \ \ \\
361\ \ \\
\underline{344}\ \ \\
172 \\
\underline{172}
\end{array}
$$

To divide two decimals when there is a decimal point in the divisor, move the decimal point to the end of the number in the divisor, and then move the decimal point the same number of places in the dividend. Place the decimal point in the answer directly above the decimal point in the dividend. Divide as usual.

EXAMPLE

Divide 33.672 ÷ 7.32

SOLUTION

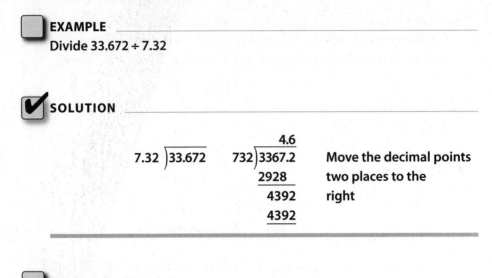

$$7.32\overline{)33.672} \qquad 732\overline{)3367.2} \qquad \text{Move the decimal points}$$

$$\begin{array}{r} 4.6 \\ 732\overline{)3367.2} \\ \underline{2928} \\ 4392 \\ \underline{4392} \end{array}$$

Move the decimal points two places to the right

TRY THESE

Perform the indicated operations

1. $63.76 + 195.2 + 3.189$

2. $195.3 - 87.215$

3. 37.3×5.6

4. $369.57 \div 97$

5. $327.6 \div 52$

SOLUTIONS

1. $\begin{array}{r} 63.760 \\ 195.200 \\ +\ \underline{3.189} \\ 262.149 \end{array}$

2. $\begin{array}{r} 195.300 \\ -\ \underline{87.215} \\ 108.085 \end{array}$

3. $\begin{array}{r} 37.3 \\ \times\ \underline{5.6} \\ 2238 \\ \underline{1865} \\ 208.88 \end{array}$

```
            3.81
4. 97)369.57
       291
       785
       776
        97
        97


        6.3
5. 52)327.6
      312
      156
      156
```

Still Struggling

Sometimes you have to add zeros to decimal numbers. Zeros can be added after the last digit on the right side of the decimal point. For example, 0.63 = 0.630 = 0.6300 = 0.63000.

This refresher reviewed how to add, subtract, multiply, and divide decimal numbers. When performing these operations, it is necessary to put the decimal point in the correct place in the answer.

Solving Word Problems Using Decimals

NOTE *If you need to review decimals, complete Refresher I.*

This section explains how to solve word problems using decimals. Many real-life problems involve decimal numbers. For example, problems involving money use decimals.

In order to solve word problems involving decimals, use the same strategies that you used in the section on operations.

EXAMPLE

If a stockbroker purchases 26 shares of stock at a cost of $8.72 per share, what is the total cost of the purchase?

SOLUTION

Goal: You are being asked to find the total cost of a stock purchase.

Strategy: Since you need to find a total and you are given two different items (dollars and shares), you multiply.

Implementation: $8.72 \times 26 = \$226.72$

Evaluation: You can check your answer using estimation: $9 \times 25 = \$225$. Since $225 is close to $226.72, the answer seems reasonable.

EXAMPLE

In 1770, the population of Maine was 31.3 thousand people, and the population of New Hampshire was 62.4 thousand people. How many more people lived in New Hampshire that year?

SOLUTION

Goal: You are being asked to find the difference in the number of people who live in two colonies.

Strategy: In order to find the difference, you need to subtract the two population values.

Implementation: $62.4 - 31.3 = 31.1$ thousand people

Evaluation: Estimate the answer by rounding 62.4 to 60 and 31.3 to 30; then subtract $60 - 30 = 30$. Since 30 is close to 31.1, the answer is probably correct.

Sometimes a word problem requires two or more steps. In this situation, you still follow the suggestions given at the beginning of this chapter to determine the operations.

EXAMPLE

Find the total cost of five picture frames at $3.59 each and two candles at $1.39 each.

✔ SOLUTION

Goal: You are being asked to find the total cost of two different items—five of one item and two of another item.

Strategy: Use multiplication to find the total cost of the picture frames and the candles, and then add the answers.

Implementation: The cost of the picture frames is 5 × $3.59 = $17.95. The cost of the candles is 2 × $1.39 = $2.78. Add the two answers: $17.95 + $2.78 = $20.73. Hence, the total cost of five picture frames and two candles is $20.73.

Evaluation: Estimate the answer: Picture frames: 5 × $3.50 = $17.50; candles: 2 × $1.40 = $2.80; total cost: $17.50 + $2.80 = $20.30. The estimated cost of $20.30 is close to the computed actual cost of $20.73; therefore, the answer is probably correct.

TRY THESE

1. The Dow Jones stock averages opened at 1,125.29 points and dropped 16.48 points. What was the closing stock average?

2. Find the cost of eight hedge trimmers if each one costs $35.75.

3. Find the total cost of an automobile trip if the person paid $156.73 for gasoline, $362.58 for lodging, $251.63 for meals, and $154.26 for miscellaneous expenses.

4. If a driver drives 261.45 miles in 6.3 hours, what is the average speed of the automobile?

5. If one kilogram weighs approximately 2.2046 pounds, what would be the approximated weight in pounds of an item that weighs 12.7 kilograms?

6. Find the cost of four outdoor chairs and two small tables if the chairs cost $17.49 each and the tables cost $19.39 each.

7. Harriet earns $10.75 per hour and gets $16.73 for each hour she works over 40 hours per week. If she works 46 hours one week, how much would she earn?

8. The weight of water is 62.5 pounds per cubic foot. Find the total weight of a tank full of water if it holds 20 cubic feet of water and the tank weighs 36.8 pounds.

9. An airport limousine service charges $15.50 plus $5.65 per mile to travel from a person's home to the airport. Find the total cost of a 12-mile trip.

10. A cell phone company charges a rate of $0.60 for the first two minutes and $0.15 for each minute after that. Find the cost of a 16-minute call.

✔ SOLUTIONS

1. $1,125.29 - 16.48 = 1,108.81$

2. $\$35.75 \times 8 = \$286.00$

3. $\$156.73 + \$362.58 + \$251.63 + \$154.26 = \$925.20$

4. $261.45 \div 6.3 = 41.5$ miles per hour

5. $2.2046 \times 12.7 = 27.99842$ pounds

6. $\$17.49 \times 4 = \$69.96$, $\$19.39 \times 2 = \$38.78$, $\$69.96 + \$38.78 = \$108.74$

7. $\$10.75 \times 40 = \$430.00$, $\$16.73 \times 6 = \$100.38$,
    $\$430.00 + \$100.38 = \$530.38$

8. $62.5 \times 20 = 1,250$ pounds, $1,250 + 36.8 = 1,286.8$ pounds

9. $\$5.65 \times 12 = \$67.80$, $\$67.80 + \$15.50 = \$83.30$

10. $\$0.60 \times 2 = \$1.20$, $\$0.15 \times 14 = \$2.10$, $\$1.20 + \$2.10 = \$3.30$

This section explained how to solve problems using decimals. Many real-life problems involve money, so it is important for you to know how to find the correct answers when decimal numbers are used.

Refresher II: Fractions

In a fraction, the top number is called the *numerator* and the bottom number is called the *denominator*.

To reduce a fraction to lowest terms, divide the numerator and denominator by the largest number that divides evenly into both numbers.

EXAMPLE

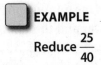
Reduce $\dfrac{25}{40}$

**SOLUTION**

$$\frac{25}{40} = \frac{25 \div 5}{40 \div 5} = \frac{5}{8}$$

To change a fraction to higher terms, divide the smaller denominator into the larger denominator, and then multiply the smaller numerator by that number to get the new numerator. This procedure will be used in addition and subtraction of fractions.

EXAMPLE

Change $\frac{5}{8}$ to 32nds

SOLUTION

Divide $32 \div 8 = 4$ and multiply $5 \times 4 = 20$. Hence, $\frac{5}{8} = \frac{20}{32}$.

This can be written as $\frac{5}{8} = \frac{5 \times 4}{8 \times 4} = \frac{20}{32}$.

An *improper fraction* is a fraction whose numerator is greater than or equal to its denominator. For example, 20/3, 6/5, and 3/3 are improper fractions. A *mixed number* is a whole number and a fraction; $8\frac{1}{3}$, $2\frac{1}{4}$, and $3\frac{5}{6}$ are mixed numbers.

To change an improper fraction to a mixed number, divide the numerator by the denominator and write the remainder as the numerator of a fraction whose denominator is the divisor. Reduce the fraction if possible.

EXAMPLE

Change $\frac{28}{6}$ to a mixed number

SOLUTION

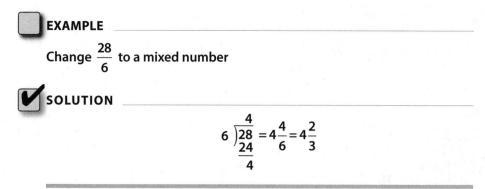

$$6\overline{)28} = 4\frac{4}{6} = 4\frac{2}{3}$$

To change a mixed number to an improper fraction, multiply the denominator of the fraction by the whole number and add the numerator. This will be the numerator of the improper fraction. Use the same number for the denominator of the improper fraction as the number in the denominator of the fraction in the mixed number.

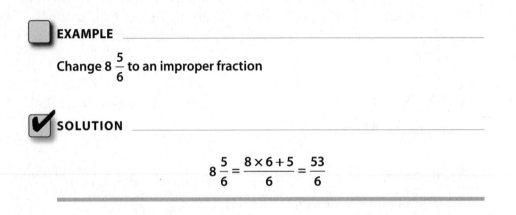

EXAMPLE

Change $8\dfrac{5}{6}$ to an improper fraction

SOLUTION

$$8\frac{5}{6} = \frac{8 \times 6 + 5}{6} = \frac{53}{6}$$

In order to add or subtract fractions, you need to find a lowest common denominator of the fractions. The *lowest common denominator* (LCD) of the fractions is the smallest number that can be divided evenly by all the denominators. For example, the LCD of 1/6, 2/3, and 7/9 is 18, since 18 can be divided evenly by 3, 6, and 9. There are several mathematical methods for finding the LCD; however, we will use the guess method. That is, just look at the denominators and figure out the LCD. If needed, you can look at an arithmetic or pre-algebra book for a mathematical method to find the LCD.

To add or subtract fractions

1. Find the LCD.

2. Change the fractions to higher terms with the LCD.

3. Add or subtract the numerators. Use the LCD.

4. Reduce or simplify the answer if possible.

EXAMPLE

Add $\dfrac{7}{8} + \dfrac{3}{10} + \dfrac{2}{5}$

SOLUTION

Use 40 as the LCD.

$$\frac{7}{8} = \frac{35}{40}$$

$$\frac{3}{10} = \frac{12}{40}$$

$$+\frac{2}{5} = \frac{16}{40}$$

$$\frac{63}{40} = 1\frac{23}{40}$$

EXAMPLE

Subtract $\dfrac{11}{12} - \dfrac{5}{9}$

SOLUTION

Use 36 as the LCD.

$$\frac{11}{12} = \frac{33}{36}$$

$$-\frac{5}{9} = \frac{20}{36}$$

$$\frac{13}{36}$$

To multiply two or more fractions, cancel if possible, multiply numerators, and then multiply denominators. Cancel means to divide out the common factors.

EXAMPLE

Multiply $\dfrac{5}{6} \times \dfrac{3}{10}$

SOLUTION

$$\frac{5}{6} \times \frac{3}{10} = \frac{\cancel{5}^{1}}{\cancel{6}_{2}} \times \frac{\cancel{3}^{1}}{\cancel{10}_{2}} = \frac{1 \times 1}{2 \times 2} = \frac{1}{4}$$

To divide two fractions, invert the fraction (turn the fraction upside down) after the ÷ sign and multiply.

EXAMPLE

Divide $\dfrac{9}{16} \div \dfrac{3}{8}$

SOLUTION

$$\frac{9}{16} \div \frac{3}{8} = \frac{9}{16} \times \frac{8}{3} = \frac{\cancel{9}^{3}}{\cancel{16}_{2}} \times \frac{\cancel{8}^{1}}{\cancel{3}_{1}} = \frac{3 \times 1}{2 \times 1} = \frac{3}{2} = 1\frac{1}{2}$$

To add mixed numbers, add the fractions, add the whole numbers, and simplify the answer if necessary.

EXAMPLE

Add $2\dfrac{4}{9} + 6\dfrac{11}{12}$

SOLUTION

$$2\frac{4}{9} = 2\frac{16}{36}$$
$$+ \ 6\frac{11}{12} = 6\frac{33}{36}$$
$$8\frac{49}{36} = 9\frac{13}{36}$$

To subtract mixed numbers, subtract the fractions, borrowing if necessary, and then subtract the whole numbers.

EXAMPLE

Subtract $15\dfrac{17}{20} - 8\dfrac{3}{5}$

SOLUTION

$$15\frac{17}{20} = 15\frac{17}{20}$$
$$- \ 8\frac{3}{5} = 8\frac{12}{20}$$
$$7\frac{5}{20} = 7\frac{1}{4}$$

(No borrowing is necessary here.)

When borrowing is necessary, take 1 away from the whole number and add it to the fraction. For example

$$8\frac{3}{5} = 8 + \frac{3}{5} = 7 + 1 + \frac{3}{5} = 7 + \frac{5}{5} + \frac{3}{5} = 7\frac{8}{5}$$

Another example:

$$11\frac{4}{9} = 11 + \frac{4}{9} = 10 + 1 + \frac{4}{9} = 10 + \frac{9}{9} + \frac{4}{9} = 10\frac{13}{9}$$

 **EXAMPLE**

Subtract $14\frac{1}{4} - 4\frac{5}{6}$

SOLUTION

$$14\frac{1}{4} = 14\frac{3}{12} = 13\frac{15}{12}$$
$$-4\frac{5}{6} = 4\frac{10}{12} = 4\frac{10}{12}$$
$$\overline{\qquad\qquad 9\frac{5}{12}}$$

To multiply or divide mixed numbers, change the mixed numbers to improper fractions, and then multiply or divide as shown previously.

EXAMPLE

Multiply $2\frac{2}{5} \times 2\frac{3}{16}$

SOLUTION

$$2\frac{2}{5} \times 2\frac{3}{16} = \frac{12}{5} \times \frac{35}{16} = \frac{\cancel{12}^{3}}{\cancel{5}_{1}} \times \frac{\cancel{35}^{7}}{\cancel{16}_{4}} = \frac{21}{4} = 5\frac{1}{4}$$

EXAMPLE

Divide $5\frac{1}{4} \div 1\frac{2}{5}$

SOLUTION

$$5\frac{1}{4} \div 1\frac{2}{5} = \frac{21}{4} \div \frac{7}{5} = \frac{\overset{3}{\cancel{21}}}{4} \times \frac{5}{\underset{1}{\cancel{7}}} = \frac{15}{4} = 3\frac{3}{4}$$

To change a fraction to a decimal, divide the numerator by the denominator.

EXAMPLE

Change $\dfrac{9}{16}$ to a decimal

SOLUTION

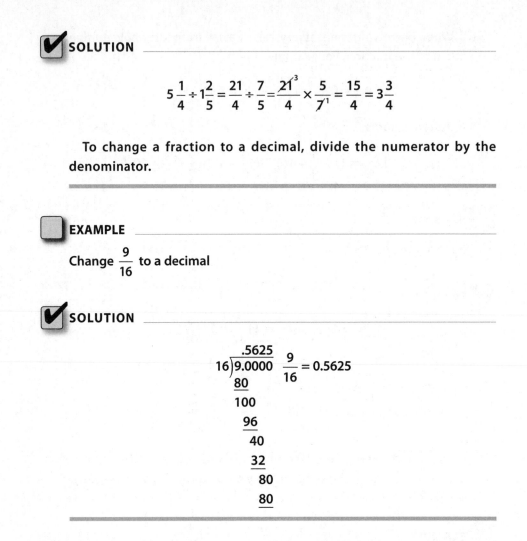

$$\begin{array}{r} .5625 \\ 16\overline{)9.0000} \\ \underline{80} \\ 100 \\ \underline{96} \\ 40 \\ \underline{32} \\ 80 \\ \underline{80} \end{array}$$

$$\frac{9}{16} = 0.5625$$

To change a decimal to a fraction, drop the decimal point and place the number over 10 if it has one decimal place, 100 if it has two decimal places, 1,000 if it has three decimal places, etc. Reduce if possible.

EXAMPLE

Change 0.88 to a fraction.

SOLUTION

$$0.88 = \frac{88}{100} = \frac{22}{25}$$

TRY THESE

1. Reduce to lowest terms: $\dfrac{18}{48}$

2. Add $\dfrac{3}{5} + \dfrac{2}{3} + \dfrac{1}{6}$

3. Subtract $\dfrac{13}{16} - \dfrac{3}{8}$

4. Multiply $\dfrac{5}{12} \times \dfrac{9}{10}$

5. Divide $\dfrac{7}{12} \div \dfrac{21}{40}$

6. Add $13\dfrac{5}{7} + 8\dfrac{4}{9}$

7. Subtract $11\dfrac{2}{9} - 7\dfrac{5}{6}$

8. Multiply $3\dfrac{1}{5} \times 6\dfrac{1}{4}$

9. Divide $4\dfrac{3}{8} \div 1\dfrac{7}{12}$

10. Change $\dfrac{13}{20}$ to a decimal.

11. Change 0.54 to a fraction.

SOLUTIONS

1. $\dfrac{18}{48} = \dfrac{18 \div 6}{48 \div 6} = \dfrac{3}{8}$

2. $\dfrac{3}{5} = \dfrac{18}{30}$

   $\dfrac{2}{3} = \dfrac{20}{30}$

   $+\dfrac{1}{6} = \dfrac{5}{30}$

   $\dfrac{43}{30} = 1\dfrac{13}{30}$

3. $\dfrac{13}{16} = \dfrac{13}{16}$

   $-\dfrac{3}{8} = \dfrac{6}{16}$

   $\dfrac{7}{16}$

4. $\dfrac{5}{12} \times \dfrac{9}{10} = \dfrac{\cancel{5}^{1}}{\cancel{12}^{4}} \times \dfrac{\cancel{9}^{3}}{\cancel{10}^{2}} = \dfrac{3}{8}$

5. $\dfrac{7}{12} \div \dfrac{21}{40} = \dfrac{\cancel{7}^{1}}{\cancel{12}^{3}} \times \dfrac{\cancel{40}^{10}}{\cancel{21}^{3}} = \dfrac{10}{9} = 1\dfrac{1}{9}$

6. $13\dfrac{5}{7} = 13\dfrac{45}{63}$

$+8\dfrac{4}{9} = 8\dfrac{28}{63}$

$\qquad 21\dfrac{73}{63} = 22\dfrac{10}{63}$

7. $11\dfrac{2}{9} = 11\dfrac{4}{18} = 10\dfrac{22}{18}$

$-7\dfrac{5}{6} = 7\dfrac{15}{18} = 7\dfrac{15}{18}$

$\qquad 3\dfrac{7}{18}$

8. $3\dfrac{1}{5} \times 6\dfrac{1}{4} = \dfrac{\cancel{16}^{4}}{\cancel{5}^{1}} \times \dfrac{\cancel{25}^{5}}{\cancel{4}^{1}} = \dfrac{20}{1} = 20$

9. $4\dfrac{3}{8} \div 1\dfrac{7}{12} = \dfrac{35}{8} \div \dfrac{19}{12} = \dfrac{35}{\cancel{8}^{2}} \times \dfrac{\cancel{12}^{3}}{19} = \dfrac{105}{38} = 2\dfrac{29}{38}$

10. $20\overline{)13.00}$ ... $\dfrac{13}{20} = 0.65$

$\begin{array}{r} .65 \\ 20\overline{)13.00} \\ \underline{120} \\ 100 \\ \underline{100} \end{array}$

11. $0.54 = \dfrac{54}{100} = \dfrac{27}{50}$

? Still Struggling

If you are having difficulties with fractions, you may need to find an arithmetic or pre-alagebra book and study the section on fractions.

This refresher reviewed the basic operations of addition, subtraction, multiplication, and division of fractions. Also, it is important to know how to change fractions to decimals and decimals to fractions.

Solving Word Problems Using Fractions

NOTE *If you need to review fractions, complete Refresher II.*

In order to solve word problems involving fractions, use the same strategies that you used in the previous sections.

EXAMPLE

A plumber is installing water pipe in a new house. He needs four pieces measuring $5\frac{5}{16}$ inches, $3\frac{1}{2}$ inches, $9\frac{3}{4}$ inches, and $11\frac{3}{8}$ inches. How long a pipe does he need to cut all the pieces from it?

 SOLUTION

Goal: You are asked to find the length of a piece of pipe necessary to cut all the pieces from it.

Strategy: Since you need to find a total and all items are in the same units (inches), use addition.

Implementation:

$$5\frac{5}{16} + 3\frac{1}{2} + 9\frac{3}{4} + 11\frac{3}{8} = 5\frac{5}{16} + 3\frac{8}{16} + 9\frac{12}{16} + 11\frac{6}{16} = 28\frac{31}{16} = 29\frac{15}{16} \text{ inches.}$$

Evaluation: You can estimate the answer since $15\frac{5}{16}$ in. is about $5\frac{1}{2}$ in., $3\frac{1}{2}$ in. can be used as is; $9\frac{3}{4}$ in. is about 10 in.; and $11\frac{3}{8}$ is about $11\frac{1}{2}$ in. Hence, $5\frac{1}{2} + 3\frac{1}{2} + 10 + 11\frac{1}{2} = 30\frac{1}{2}$. Since $30\frac{1}{2}$ in. is close to 31 in., your answer is probably correct.

EXAMPLE

A bus travels $75\frac{3}{8}$ miles in $2\frac{1}{4}$ hours. What is the average speed of the bus?

✅ SOLUTION

Goal: You are asked to find the average speed of the bus.

Strategy: Since you are given a total distance and the time it took, you divide the total distance by the time to get the average speed.

Implementation: $75\dfrac{3}{8} \div 2\dfrac{1}{4} = \dfrac{603}{8} \div \dfrac{9}{4} = \dfrac{\cancel{603}^{\,201}_{}{}^{67}}{\cancel{8}_2} \times \dfrac{\cancel{4}^{1}}{\cancel{9}^{3}{}^{1}} = \dfrac{67}{2} = 33\dfrac{1}{2}.$

Hence, the average speed is $33\dfrac{1}{2}$ miles per hour.

Evaluation: You can check by multiplying $2\dfrac{1}{4} \times 33\dfrac{1}{2} = 75\dfrac{3}{8}$

EXAMPLE

If the Tigers are $3\dfrac{1}{2}$ games behind the Cougars in the baseball standings and the Wildcats are 5 games behind the Cougars, how many games are the Wildcats behind the Tigers?

✅ SOLUTION

Goal: You are asked to find how many games the Wildcats are behind the Tigers.

Strategy: Since you need to find how many games behind the Wildcats are, you use subtraction.

Implementation: $5 - 3\dfrac{1}{2} = 4\dfrac{2}{2} - 3\dfrac{1}{2} = 1\dfrac{1}{2}$ games. Hence the Wildcats are $1\dfrac{1}{2}$ games behind the Tigers in the standings.

Evaluation: You can check the solution by adding $3\dfrac{1}{2} + 1\dfrac{1}{2} = 5$

TRY THESE

1. A tradesman can assemble a cable pulley system in $5\dfrac{1}{8}$ hours while his assistant can do the same job in $7\dfrac{2}{3}$ hour. How much faster can the tradesman do the job?

2. One cubic foot of oil is about $7\dfrac{1}{2}$ gallons. How many cubic feet of oil would a 20-gallon container hold?

3. Joanne worked $6\dfrac{1}{2}$ hours on Monday, $4\dfrac{3}{4}$ hours on Tuesday, $5\dfrac{1}{8}$ hours on Wednesday, and 3 hours on Thursday. Find the total number of hours she worked that week.

4. A train travels from Pittsburgh to Chicago in $12\frac{5}{8}$ hours while another train made the same trip in $9\frac{5}{6}$ hours. How much faster was the second train?

5. How many identification cards that are $1\frac{5}{8}$ inches long can be cut from a piece of card stock $17\frac{7}{8}$ inches long?

6. The scale on a map states that $\frac{3}{4}$ inch is equal to 20 miles. Find the distance in miles between two towns if it measures $4\frac{7}{8}$ inches on the map.

7. Jeanne cut three pieces of ribbon that measured $5\frac{2}{3}$ inches, $7\frac{5}{12}$ inches, and $4\frac{1}{4}$ inches long. If the total length of the ribbon is $21\frac{7}{8}$ inches long, how much of the ribbon was left?

8. Eugene purchased a laptop computer for $800. He made a down payment of $\frac{1}{5}$ of the price and paid the balance in eight monthly installments. How much did he pay each month?

9. If a $264,000 home is assessed at $\frac{2}{3}$ of its value, find the assessed value of the house.

10. Find the distance around a triangular piece of property if the sides measure $106\frac{3}{8}$ feet, $96\frac{1}{6}$ feet, and $101\frac{2}{3}$ feet.

✔ **SOLUTIONS**

1. $7\frac{2}{3} - 5\frac{1}{8} = 7\frac{16}{24} - 5\frac{3}{24} = 2\frac{13}{24}$ hours

2. $20 \div 7\frac{1}{2} = \frac{20}{1} \div \frac{15}{2} = \frac{\overset{4}{\cancel{20}}}{1} \times \frac{2}{\underset{3}{\cancel{15}}} = \frac{8}{3} = 2\frac{2}{3}$ cubic feet

3. $6\frac{1}{2} + 4\frac{3}{4} + 5\frac{1}{8} + 3 = 6\frac{4}{8} + 4\frac{6}{8} + 5\frac{1}{8} + 3 = 18\frac{11}{8} = 19\frac{3}{8}$ hours

4. $12\frac{5}{8} - 9\frac{5}{6} = 12\frac{15}{24} - 9\frac{20}{24} = 11\frac{39}{24} - 9\frac{20}{24} = 2\frac{19}{24}$ hours

5. $17\frac{7}{8} \div 1\frac{5}{8} = \frac{143}{8} \div \frac{13}{8} = \frac{\overset{11}{\cancel{143}}}{\cancel{8}^{1}} \times \frac{\cancel{8}^{1}}{\cancel{13}^{1}} = \frac{11}{1} = 11$ cards

6. $4\frac{7}{8} \div \frac{3}{4} = \frac{\overset{13}{\cancel{39}}}{\cancel{8}^{2}} \times \frac{\cancel{4}^{1}}{\cancel{3}^{1}} = \frac{13}{2} = 6\frac{1}{2}$ inches; $6\frac{1}{2} \times 20 = \frac{13}{\cancel{2}^{1}} \times \frac{\overset{10}{\cancel{20}}}{1} = 130$ miles

7. $5\dfrac{2}{3} + 7\dfrac{5}{12} + 4\dfrac{1}{4} = 5\dfrac{8}{12} + 7\dfrac{5}{12} + 4\dfrac{3}{12} = 16\dfrac{16}{12} = 17\dfrac{1}{3};$

$21\dfrac{7}{8} - 17\dfrac{1}{3} = 21\dfrac{21}{24} - 17\dfrac{8}{24} = 4\dfrac{13}{24}$ inches

8. $\dfrac{1}{5} \times \$800 = \dfrac{1}{\cancel{5}^{1}} \times \dfrac{\cancel{800}^{\,160}}{1} = \$160;\ \$800 - \$160 = \$640;\ \$640 \div 8 = \$80$

9. $\dfrac{2}{3} \times \$264,000 = \dfrac{2}{\cancel{3}^{1}} \times \dfrac{\cancel{264,000}^{\,88,000}}{1} = \$176,000$

10. $106\dfrac{3}{8} + 96\dfrac{1}{6} + 101\dfrac{2}{3} = 106\dfrac{9}{24} + 96\dfrac{4}{24} + 101\dfrac{16}{24} = 303\dfrac{29}{24} = 304\dfrac{5}{24}$ feet

In this section, you learned how to solve word problems using fractions.

Summary

Chapter 2 explained the important words and concepts that will enable you to determine which operations (addition, subtraction, multiplication, or division) to use when solving word problems using whole numbers, decimals, or fractions.

QUIZ

1. The island of Puerto Rico contains 3,339 square miles, while the island of Jamaica contains 4,244 square miles. How much larger is the island of Jamaica?
   A. 7,583 square miles
   B. 905 square miles
   C. 6,354 square miles
   D. 1,003 square miles

2. Find the cost of 8 feet of ribbon if it sells for $1.59 per foot.
   A. $12.72
   B. $14.52
   C. $ 9.32
   D. $16.82

3. The length of Lake Superior is 350 miles. The length of Lake Huron is 206 miles, and the length of Lake Erie is 241 miles. Find the total length of all three lakes.
   A. 834 miles
   B. 973 miles
   C. 797 miles
   D. 743 miles

4. If a person earns $66,000 a year, what is the person's monthly salary?
   A. $4,000
   B. $4,200
   C. $4,500
   D. $5,500

5. A carpenter made six shelves that were $2\frac{3}{4}$ feet long and three shelves that were $5\frac{1}{8}$ feet long. How much lumber did he use?

   A. $31\frac{7}{8}$ feet

   B. $70\frac{7}{8}$ feet

   C. $39\frac{3}{8}$ feet

   D. $54\frac{1}{8}$ feet

6. A professor said $\frac{5}{6}$ of his students are juniors. If there are 72 students in his classes, how many of them are juniors?

    A. 24
    B. 56
    C. 48
    D. 60

7. To change a Fahrenheit temperature to a Celsius temperature, subtract 32°, and then take $\frac{5}{9}$ of the answer. What is the Celsius temperature for a Fahrenheit reading of 86°?

    A. $79\frac{7}{9}°$
    B. 30°
    C. $28\frac{2}{5}°$
    D. 44°

8. A person purchased a digital camera for $25 down and eight monthly payments of $16.65. Find the total cost of the camera.

    A. $41.55
    B. $49.55
    C. $158.20
    D. $216.55

9. If 4 servings of a recipe call for $1\frac{3}{4}$ cups of flour, how much flour will be needed to make 12 servings?

    A. 21 cups
    B. 7 cups
    C. $2\frac{7}{8}$ cups
    D. $5\frac{1}{4}$ cups

10. A person made the following purchases: $18.77, $42.56, $51.75, and $14.36. Find the total amount spent.

    A. $132.59
    B. $127.44
    C. $155.62
    D. $142.73

In **New Jersey**, government owns **129,79** ~~cres~~ of land. **In Texas,** ~~~ment~~ ~~own~~ **2,307,171** ~~acres~~ land and in ~~lan~~ go~~ernment~~ ~~213~~ of land. **Find** the total amount of land own~~ed~~ y the federal go~~ernment~~ in the ~~three~~ states.

chapter **3**

Solving Percent Problems

This chapter reviews the concept of percent and the three types of percent problems. Finally, word problems using percents are explained.

CHAPTER OBJECTIVES

In this chapter, you will

- Review how to change percents to decimals, change decimals to percents, change percents to fractions, change fractions to percents, and solve the three types of percent problems
- Learn how to solve word problems using percents

Refresher III: Percents

Percent means hundredths or part of a hundred. For example, 42% means 0.42 or 42/100. You can think of 42% as a square being divided into 100 equal parts and 42% is 42 equal parts out of 100 equal parts.

To change a percent to a decimal, drop the % sign and move the decimal point two places to the left. The decimal point in 42% is between the 2 and the % sign. It is not written.

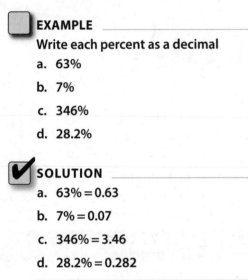

EXAMPLE

Write each percent as a decimal

a. 63%

b. 7%

c. 346%

d. 28.2%

SOLUTION

a. 63% = 0.63

b. 7% = 0.07

c. 346% = 3.46

d. 28.2% = 0.282

To change a decimal to a percent, move the decimal two places to the right and affix the percent sign.

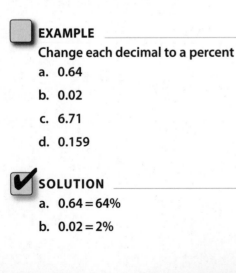

EXAMPLE

Change each decimal to a percent

a. 0.64

b. 0.02

c. 6.71

d. 0.159

SOLUTION

a. 0.64 = 64%

b. 0.02 = 2%

    c. $6.71 = 671\%$

    d. $0.159 = 15.9\%$

To change a percent to a fraction, drop the percent sign and place the number in the numerator of a fraction whose denominator is 100. Reduce or simplify if necessary.

EXAMPLE

Change each percent to a fraction

  a. 80%

  b. 55%

  c. 175%

  d. 5%

SOLUTION

  a. $80\% = \dfrac{80}{100} = \dfrac{4}{5}$

  b. $55\% = \dfrac{55}{100} = \dfrac{11}{20}$

  c. $175\% = \dfrac{175}{100} = 1\dfrac{3}{4}$

  d. $5\% = \dfrac{5}{100} = \dfrac{1}{20}$

To change a fraction to a percent, change the fraction to a decimal and then change the decimal to a percent.

EXAMPLE

Change each fraction or mixed number to a percent

  a. $\dfrac{4}{5}$

  b. $\dfrac{7}{10}$

  c. $\dfrac{3}{8}$

  d. $2\dfrac{3}{4}$

✔ SOLUTION

a. $5\overline{)4.0}$ ... $\dfrac{4}{5} = 0.8 = 80\%$

$\quad\quad\;\; .8$
$\quad\quad\;\; \underline{40}$
$\quad\quad\quad\; 0$

b. $10\overline{)7.0}$ ... $\dfrac{7}{10} = 0.7 = 70\%$

$\quad\quad\;\; .7$
$\quad\quad\;\; \underline{70}$
$\quad\quad\quad\; 0$

c. $8\overline{)3.000}$ ... $\dfrac{3}{8} = 0.375 = 37.5\%$

$\quad\quad\;\; .375$
$\quad\quad\;\; \underline{24}$
$\quad\quad\quad 60$
$\quad\quad\quad \underline{56}$
$\quad\quad\quad\;\; 40$
$\quad\quad\quad\;\; \underline{40}$
$\quad\quad\quad\quad 0$

d. $4\overline{)11.00}$ ... $2\dfrac{3}{4} = \dfrac{11}{4} = 275\%$

$\quad\quad\;\; 2.75$
$\quad\quad\;\; \underline{8}$
$\quad\quad\;\; 30$
$\quad\quad\;\; \underline{28}$
$\quad\quad\;\; 20$
$\quad\quad\;\; \underline{20}$
$\quad\quad\quad 0$

A percent word problem has three numbers—the whole, total, or base (B); the part (P); and the rate or percent (R). Suppose that in a class of 25 students, 8 are absent. Now the whole or total is 25 and the part is 8. The rate or percent of students who were absent is $8/25 = 0.32 = 32\%$.

In a percent problem, you will be given two of the three numbers and will be asked to find the third number. Percent problems can be solved by using a percent circle. The circle is shown in Figure 3-1.

In the top portion of the circle, write the word part (P). In the lower right portion of the circle, write the word rate (R), and in the lower left portion, write the word base (B). Put a multiplication sign between the two lower portions and a division sign between the top and bottom portions.

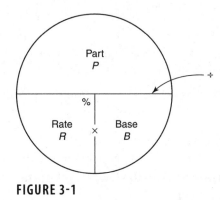

FIGURE 3-1

If you are asked to find the part (P), place the rate (R) in the lower left portion of the circle and the base (B) in the lower right portion. The circle tells you to use the formula $P = R \times B$ and multiply.

If you are asked to find the rate (R), place the part (P) in the top portion of the circle and the base (B) in the lower right portion. The circle tells you to use the formula $R = P/B$ and divide. The answer will be in decimal form and needs to be changed to a percent.

If you are asked to find the base, place the part (P) in the top portion and the rate (R) in the bottom left portion. The circle tells you to use the formula $B = P/R$ and divide. See Figure 3-2.

Still Struggling

Be sure to change the percent to a decimal or fraction before multiplying or dividing.

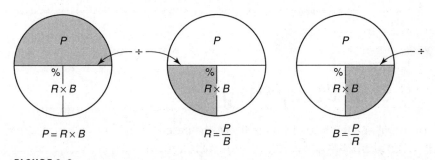

FIGURE 3-2

Type I: Finding the Part

 EXAMPLE
Find 42% of 36

SOLUTION
Since 42% is the rate, place it in the lower left portion of the circle, and since 36 is the base, place it in the lower right portion of the circle and then multiply. See Figure 3-3.

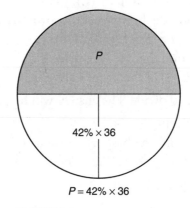

$$P = 42\% \times 36$$

FIGURE 3-3

$$P = R \times P$$
$$= 42\% \times 36$$
$$= 0.42 \times 36$$
$$= 15.12$$

? Still Struggling

The number after the word "of" is always the base.

Type II: Finding the Rate

EXAMPLE

16 is what percent of 20?

SOLUTION

Since 16 is the part, place it in the top portion of the circle, and since 20 is the base, place it in the lower right portion of the circle, and then divide. See Figure 3-4.

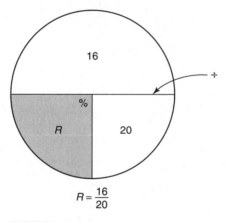

$$R = \frac{16}{20}$$

FIGURE 3-4

$$R = \frac{P}{B}$$

$$R = \frac{16}{20}$$

$$= 0.80 = 80\%$$

Type III: Finding the Base

EXAMPLE

48 is 60% of what number?

SOLUTION

Since 48 is the part, place it in the top portion of the circle, and since 60% is the rate, place it in the lower right portion of the circle and then divide. See Figure 3-5.

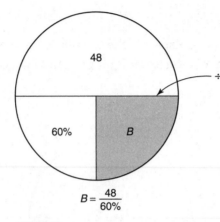

$$B = \frac{48}{60\%}$$

FIGURE 3-5

$$B = \frac{P}{R}$$

$$= \frac{48}{60\%}$$

$$= \frac{48}{0.60}$$

$$= 80$$

TRY THESE

1. What percent of 60 is 45?

2. Find 13% of 37.

3. What percent of 64 is 48?

4. 150 is 25% of what number?

5. Find 84% of 63.

6. 72 is 24% of what number?

7. What percent of 35 is 21?

8. 16 is what percent of 40?

9. 15 is what percent of 60?

10. Find 15% of 90.

✔ SOLUTIONS

1. $R = \dfrac{P}{B}$

    $= \dfrac{45}{60}$

    $= 0.75 = 75\%$

2. $P = R \times B$

    $= 13\% \times 37$

    $= 0.13 \times 37$

    $= 4.81$

3. $R = \dfrac{P}{B}$

    $= \dfrac{48}{64}$

    $= 0.75 = 75\%$

4. $B = \dfrac{P}{R}$

    $= \dfrac{150}{25\%}$

    $= \dfrac{150}{0.25}$

    $= 600$

5. $P = R \times B$

    $= 84\% \times 63$

    $= 0.84 \times 63 = 52.92$

6. $B = \dfrac{P}{R}$

    $= \dfrac{72}{24\%}$

    $= \dfrac{72}{0.24}$

    $= 300$

7. $R = \dfrac{P}{B}$

$= \dfrac{21}{35}$

$= 0.60 = 60\%$

8. $R = \dfrac{P}{B}$

$= \dfrac{16}{40}$

$= 0.4 = 40\%$

9. $R = \dfrac{P}{B}$

$= \dfrac{15}{60}$

$= 0.25 = 25\%$

10. $P = R \times B$

$= 15\% \times 90$

$= 0.15 \times 90$

$= 13.5$

? Still Struggling

Remember: The number after the word "of" is always the base, the number with the percent sign (%) is always the rate, and the number immediately preceding or following the word "is" is the part.

In this refresher, you have reviewed how to convert among percents, decimals, and fractions. There are three basic types of percent problems. They use the base, the rate, and the part. You will be given two numbers and be asked to find the third number.

Solving Percent Word Problems

NOTE *If you need to review percents, complete Refresher III.*

A percent problem consists of three values, the base, the rate, and the part. The base
(*B*) is the whole or total, and the rate (*R*) is a percent. One of these three will be
unknown. For example, if a box contains 10 calculators, then the whole is 10. If four
calculators are placed on a store's shelf, then 4 is the part. Finally, the percent is
4/10 = 0.40 = 40%. That is, 40% of the calculators were placed on the store's shelf.

Percent problems can be solved using the circle method. Figure 3-6 shows
how to use the circle method to solve percent problems.

In the top of the circle, place the part (*P*). In the lower left portion of the
circle, place the rate (%), and in the lower right portion, place the base (*B*).
Now if you are given the bottom two numbers, multiply. That is, $P = R \times B$. If
you are given the top number, the part, and one of the bottom numbers, divide
to find the other number. That is, $R = P/B$ or $B = P/R$. See Figure 3-7.

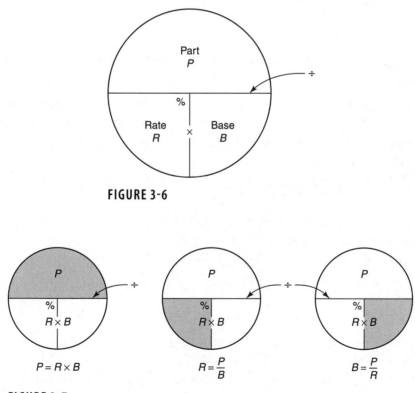

FIGURE 3-6

FIGURE 3-7

There are three types of percent word problems. They are

Type I: Finding the part

Type II: Finding the rate

Type III: Finding the base

In order to solve percent problems, read the problem and identify the base, rate, and part. One of the three will be unknown. Substitute the two known quantities in the circle and use the correct formula to find the unknown value. Be sure to change the percent to a decimal before multiplying or dividing.

Type I: Finding the Part

In Type I problems, you are given the base and rate and you are asked to find the part.

EXAMPLE

There are 40 preowned automobiles on a lot. If 30% of them are white, how many of the automobiles are white?

SOLUTION

Goal: You are being asked to find the number of automobiles that are white.

Strategy: Draw the circle and place 30% in the lower left portion of the circle and 40 in the lower right portion of the circle since it is the total number of automobiles in the lot. To find the part, use $P = R \times B$. See Figure 3-8.

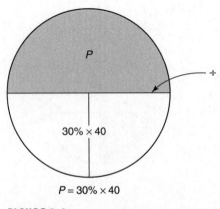

$$P = 30\% \times 40$$

FIGURE 3-8

Implementation: Substitute in the formula and solve for *P*.

$$P = R \times B$$

$$P = 30\% \times 40$$

$$P = 0.30 \times 40$$

$$P = 12 \text{ automobiles}$$

Hence, 12 automobiles are white.

Evaluation: Since $30\% = \dfrac{3}{10}$ and $\dfrac{3}{10}$ of $40 = 12$, the answer is correct.

Type II: Finding the Rate (%)

In Type II problems, you are given the part and the whole and you are asked to find the rate as a percent. The answer obtained from the formula will be in decimal form. Make sure that you change it into a percent.

EXAMPLE

A person bought a textbook for $35 and paid a sales tax of $2.10. Find the tax rate.

SOLUTION

Goal: You are being asked to find the rate (%).

Strategy: In this case, the base (*B*) is the total cost, which is $35, and the sales tax, $2.10, is the part. Draw the circle and put $35 in the lower right portion and $2.10 in the top portion. To find the rate, use $R = \dfrac{P}{B}$. See Figure 3-9.

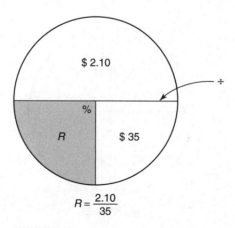

$$R = \frac{2.10}{35}$$

FIGURE 3-9

Implementation:

$$R = \frac{P}{B}$$

$$R = \frac{\$2.10}{\$35}$$

$$R = 0.06 = 6\%$$

The sales tax rate is 6%.

Evaluation: **To check the answer, find 6% of $35: 6% × 35 = 0.06 × 35 = $2.10. The answer is correct.**

Type III: Finding the Base

In Type III problems, you are given the part and rate and are asked to find the base or whole.

EXAMPLE

A salesperson earns a 15% commission on all sales. If the commission was $2435.25, find the amount of his sales.

SOLUTION

Goal: You are being asked to find the total amount of sales.

Strategy: In this type of problem, you are given the part (commission) and the rate. Place $2435.25 in the top portion of the circle and the 15% in the bottom left portion. Use $B = \frac{P}{R}$. See Figure 3-10.

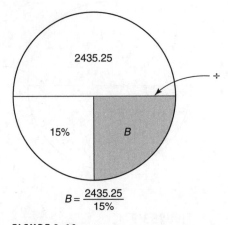

$$B = \frac{2435.25}{15\%}$$

FIGURE 3-10

Implementation:

$$B = \frac{P}{R}$$

$$B = \frac{\$2435.25}{15\%} = \frac{2435.25}{0.15} = \$16,235$$

The total sales were $16,235.

Evaluation: To check the answer, find 15% of $16,235: 0.15 × $16,235 = $2435.25. Hence, the answer is correct.

Some percent problems involve finding a percent increase or decrease. Always remember that the original value is used as the base and the amount of the increase or decrease is used as the part. For example, suppose an alarm clock sold for $50 last week and is on sale for $40 this week. The decrease is $50 − $40 = $10. The percent of decrease is 10/50 = 0.20 or 20%.

EXAMPLE

Willis increased the fiber content of his diet from 12 to 15 grams a day. Find the percent of increase in the daily fiber.

SOLUTION

Goal: You are being asked to find the percent of the increase in the amount of fiber he consumed.

Strategy: Find the increase, and then place that number in the top portion of the circle. The base is the original amount. Use $R = \dfrac{P}{B}$. See Figure 3-11.

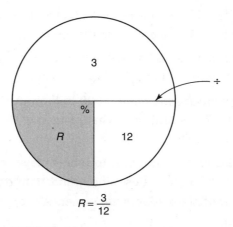

FIGURE 3-11

Implementation: The increase is $15 - 12 = 3$ grams.

$$R = \frac{P}{B}$$

$$R = \frac{3}{12} = \frac{1}{4} = 0.25 = 25\%$$

Hence, the rate of increase is 25%.

Evaluation: Find 25% of 12: $0.25 \times 12 = 3$. The solution is correct.

TRY THESE

1. Cindy earns a 17% commission on all the sales she makes. What was her commission on a $5,320 sale?

2. A quarterback completes 40% of his passes in a game. If he completed 12 passes, how many did he throw?

3. If 3 gallons of oil are removed from a full 20-gallon tank, what percent of the oil remains in the tank?

4. A tire is on sale for $120. If it is purchased when the sale price is 20% off the original price, what was the original price?

5. Find the rate of the sales tax if the tax on an item costing $32.60 is $1.63.

6. On a 60-point exam, Sam missed 9 questions. What was his percent score?

7. If the sale price of a desk was $432, and the sale price discount was 20% off the original price, find the original price.

8. If 30% of Internet users use English, in a survey of 1,600 people, find the total number of people who used English on the Internet.

9. If the movie *Titanic* made $600 million in the United States and $1,848 million worldwide, what percent of the income was made in the United States?

10. Jean Borotra, a championship tennis player from France, won four grand slam singles titles, nine grand slam doubles titles, and five mixed grand slam titles in her career. What percent of her wins were doubles?

SOLUTIONS

1. $P = R \times B$

    $= 17\% \times \$5320 = 0.17 \times \$5320 = \$904.40$

2. $R = \dfrac{P}{B}$

    $= \dfrac{12}{40\%} = \dfrac{12}{0.40} = 30$

3. $R = \dfrac{P}{B}$

    $= \dfrac{3}{20} = 0.15 = 15\%: 100\% - 15\% = 85\%$

4. $100\% - 20\% = 80\%$ (The sale price is 80% of the original price.)

    $B = \dfrac{P}{B}$

    $= \dfrac{\$120}{80\%} = \dfrac{\$120}{0.80} = \$150$

5. $R = \dfrac{P}{B}$

    $= \dfrac{\$1.63}{\$32.60} = 0.05 = 5\%$

6. $60 - 6 = 54$ (Sam got 54 out of 60 answers correct.)

    $B = \dfrac{P}{B}$

    $= \dfrac{54}{60} = 0.9 = 90\%$

7. $100\% - 20\% = 80\%$ (The sale price is 80% of the original price.)

    $B = \dfrac{P}{B}$

    $= \dfrac{\$432}{80\%} = \dfrac{\$432}{0.8} = \$540$

8. $P = R \times B$

    $= 30\% \times 1600 = 0.3 \times 1600 = 480$

9. $R = \dfrac{P}{B}$

    $= \dfrac{600}{1,848} = 0.325 \text{ (rounded)} = 32.5\%$

10. The total number of grand slam victories is $4 + 9 + 5 = 18$

$$R = \frac{P}{B}$$

$$= \frac{9}{18} = 0.5 = 50\%$$

In this section, you have learned to solve the basic types of percent problems. There are three formulas that are used. They are $P = B \times R$; $R = P/B$; and $B = P/R$.

Summary

In this chapter, the basic concepts of percent were explained. The basic conversions of percents to decimals or fractions were shown. The conversions of decimals or fractions to percents were also explained. Finally, the solutions to the three basic types of percent problems were shown.

QUIZ

1. Find the sales tax on a lounge chair that costs $39 if the rate is 6%.
   A. $1.95
   B. $0.24
   C. $2.34
   D. $1.95

2. A salesperson received a commission of $80 on a sale of an item. If his commission rate is 16%, find the amount of the sale.
   A. $13
   B. $12
   C. $500
   D. $12.80

3. If a person earned $48,000 a year and received a $1,200 raise, what was the percent increase in her salary?
   A. 25%
   B. 0.25%
   C. 2.5%
   D. 250%

4. If a family purchased a home for $160,000 and put 18% down, how much was left to finance?
   A. $28,800
   B. $128,000
   C. $131,200
   D. $32,000

5. If a calculator originally sold for $60 and was reduced 25% for a sale, what was the reduced price?
   A. $40
   B. $45
   C. $25
   D. $15

6. Mary took a 40-problem mathematics quiz. If she received a grade of 85%, how many problems did she miss?
   A. 34
   B. 32
   C. 18
   D. 6

7. A department store has 72 employees. On a very snowy day, there were 18 employees absent. What percent of the employees were absent?
   A. 25%
   B. 36%
   C. 75%
   D. 84%

8. Frank earned $1,800 per month. If he received a 6% salary increase, how much does he earn now?
   A. $108
   B. $1908
   C. $2118
   D. $96

9. A certain mixture of peanuts and cashews consists of 32% cashews. If the total weight of the mixture is 50 pounds, how many pounds of the mixture consists of peanuts?
   A. 32 pounds
   B. 16 pounds
   C. 64 pounds
   D. 34 pounds

10. A person bought a necklace for $800. If she made a down payment of $250 and paid the balance in 11 monthly installments, how much did she pay each month?
   A. $50.00
   B. $95.45
   C. $550.00
   D. $65.00

chapter 4

Solving Proportion and Formula Problems

This chapter explains how to solve word problems using proportions and how to evaluate formulas. Many real-world problems can be solved by using these two techniques.

CHAPTER OBJECTIVES

In this chapter, you will learn how to

- Solve word problems using proportions
- Solve word problems using formulas

Ratios

In order to solve word problems using proportions, it is necessary to understand the concept of a ratio. A *ratio* is a comparison of two numbers by using division. For example, the ratio of 6 to 10 is 6/10, which reduces to 3/5.

Ratios are used to make comparisons between quantities. If you drive 180 miles in 4 hours, then the ratio of miles to hours is 180/4 or 45/1. In other words, you averaged 45 miles per hour.

It is important to understand that whatever number comes first in a ratio statement is placed in the numerator of the fraction and whatever number comes second in the ratio statement is placed in the denominator of the fraction. In general, the ratio of *a* to *b* is written as *a/b*. Ratios can be written with a colon. For example, the ratio of 3 to 5 can be written as 3:5.

Proportions

A *proportion* is a statement of equality of two ratios. For example, 4/5 = 20/25 is a proportion. Proportions can also be written using a colon. For example, the proportion 4/5 = 20/25 can be written as 4:5 = 20:25 or 4:5::20:25.

A proportion consists of four terms, and it is usually necessary to find one of the terms of the proportion given the other three terms. This can be done by cross-multiplying and then dividing both sides of the equation by the numerical coefficient of the variable.

EXAMPLE

Find the value of $\dfrac{5}{7} = \dfrac{x}{35}$

SOLUTION

$$\frac{5}{7} = \frac{x}{35}$$

$7 \cdot x = 5 \cdot 35$ **Cross-multiply**

$7x = 175$

$\dfrac{\cancel{7}^{1} x}{\cancel{7}^{1}} = \dfrac{175}{7}$ **Divide by 7**

$x = 25$

EXAMPLE

Find the value of x: $\dfrac{12}{x} = \dfrac{36}{6}$

SOLUTION

$$\frac{12}{x} = \frac{36}{6}$$

$$36 \cdot x = 12 \cdot 6 \qquad \text{Cross-multiply}$$

$$36x = 72$$

$$\frac{\cancel{36}^{1} x}{\cancel{36}_{1}} = \frac{72}{36} \qquad \text{Divide by 36}$$

$$x = 2$$

The strategy used to solve problems involving proportions is to identify and write the ratio statement and then write the proportion. Let the unknown term be x; then cross-multiply and solve for x.

EXAMPLE

If a person burns 110 calories in 8 minutes of running, how many calories will the person burn if she runs for 30 minutes?

SOLUTION

Goal: You are being asked to determine how many calories can be burned in 30 minutes of running.

Strategy: Write the ratio, and then set up the proportion. The ratio statement is 110 calories to 8 minutes or $\dfrac{110\,\text{calories}}{8\,\text{minutes}}$. Set up the proportion. It is $\dfrac{110\,\text{calories}}{8\,\text{minutes}} = \dfrac{x\,\text{calories}}{30\,\text{minutes}}$.

Implementation: Solve the proportion.

$$\frac{110 \text{ calories}}{8 \text{ minutes}} = \frac{x \text{ calories}}{30 \text{ minutes}}$$

$$8x = 110 \cdot 30$$

$$8x = 3{,}300$$

$$\frac{\cancel{8}^{1} x}{\cancel{8}^{1}} = \frac{3,300}{8}$$

$$x = 412.5 \text{ calories}$$

The runner will burn 412.5 calories if she runs for 30 minutes.

Evaluation:

$$\frac{110}{8} = \frac{412.5}{30}$$

Substitute the value for x in the proportion and cross-multiply. If the cross-products are equal, the answer is correct.

$$30 \cdot 110 = 8 \cdot 412.5$$

$$3,300 = 3,300$$

? Still Struggling

Notice that when you set up a proportion, always place the same units in the numerators and the same units in the denominators. In the previous problem, $\frac{\text{calories}}{\text{minutes}} = \frac{\text{calories}}{\text{minutes}}$.

EXAMPLE

If a grocery store sells canned pears at a price of 4 for $10, what is the cost of 10 cans?

SOLUTION

Goal: You are being asked to find the cost of 10 cans of pears.

Strategy: Write the ratio. It is $\frac{4 \text{ cans}}{\$10}$. The proportion is $\frac{4 \text{ cans}}{\$10} = \frac{10 \text{ cans}}{x}$.

Implementation: Solve the proportion.

$$\frac{4 \text{ cans}}{\$10} = \frac{10 \text{ cans}}{x}$$

$$4x = 10 \cdot 10$$

$$4x = 100$$

$$\frac{\cancel{4}^1 x}{\cancel{4}^1} = \frac{100}{4}$$

$$x = \$25$$

Evaluation:

$$\frac{4}{\$10} = \frac{10}{\$25}$$

$$4 \cdot \$25 = 10 \cdot \$10$$

$$\$100 = \$100$$

The answer is correct.

EXAMPLE

If four gallons of paint can cover 1,240 square feet, how many square feet will seven gallons of paint cover?

SOLUTION

Goal: You are being asked to find how many square feet seven gallons of paint will cover.

Strategy: Write the ratio. It is $\frac{4\,\text{gallons}}{1,240\,\text{sq ft}}$. The proportion is $\frac{4\,\text{gallons}}{1,240\,\text{sq ft}} = \frac{7\,\text{gallons}}{x\,\text{sq ft}}$.

Implementation: Solve the proportion.

$$\frac{4\,\text{gallons}}{1,240\,\text{sq ft}} = \frac{7\,\text{gallons}}{x\,\text{sq ft}}$$

$$7 \cdot 1,240 = 4x$$

$$8,680 = 4x$$

$$\frac{8,680}{4} = \frac{\cancel{4}^1 x}{\cancel{4}^1}$$

$$2,170 \text{ square feet} = x$$

Seven gallons will cover 2,170 square feet.

Evaluation:

$$\frac{4}{1,240} = \frac{7}{2,170}$$

$$1,240 \cdot 7 = 4 \cdot 2,170$$

$$8,680 = 8,680$$

The answer is correct.

EXAMPLE

If a tree casts a shadow of 10 feet and a 6-foot pole casts a shadow of 3.2 feet, how tall is the tree?

SOLUTION

Goal: You are being asked to find the height of the tree.

Strategy: The ratio statement is $\dfrac{6\,\text{feet}}{3.2\,\text{feet}}$. The proportion is $\dfrac{6\,\text{feet}}{3.2\,\text{feet}} = \dfrac{x\,\text{feet}}{10\,\text{feet}}$.

Implementation:

$$\frac{6}{3.2} = \frac{x}{10}$$

$$6 \cdot 10 = 3.2x$$

$$60 = 3.2x$$

$$\frac{60}{3.2} = \frac{\overset{1}{\cancel{3.2}}\,x}{\underset{1}{\cancel{3.2}}}$$

$$18.75\,\text{feet} = x$$

The tree is 18.75 feet tall.

Evaluation:

$$\frac{6}{3.2} = \frac{18.75}{10}$$

$$3.2 \cdot 18.75 = 6 \cdot 10$$

$$60 = 60$$

The answer is correct.

Still Struggling

As long as you keep the same units in the numerators and the same units in the denominators of a proportion, it does not matter how the proportion is set up. For example, the proportion $x/10 = 6/5$ will give the same answer for x as the proportion $10/x = 5/6$.

TRY THESE

1. On a map, the scale is $\frac{3}{4}$ inch = 150 miles. How far apart are two cities whose distance on a map is $4\frac{1}{2}$ inches?

2. If a recipe calls for 2.4 cups of flour and it serves six people, how many cups of flour will be needed to serve two people?

3. If a merchant can order 12 shirts for $280, how much will 15 shirts cost?

4. If three pounds of grass seed will cover 1,320 square feet, how many pounds will be needed to cover 3,080 square feet?

5. If a person drives an automobile 8,100 miles in 9 months, about how many miles will the person drive the automobile in 15 months?

6. Sam wants to waterproof his patio deck. If two gallons of sealant can cover 500 square feet, how many gallons should he buy if his deck is 1,460 square feet?

7. Betty can bicycle 216 miles in 9 days; how far can she travel in 14 days?

8. If an author can write two chapters in 11 days, how many days will it take her to complete a 16-chapter book?

9. Sophia wants to save money for a trip. If she calculates that she can save $456 in three months, how many months will it take her to save for a trip costing $3,648?

10. If Abigail can purchase six concert tickets for $112.50, how many tickets can she purchase for $168.75?

☑ SOLUTIONS _____

1. $\dfrac{\dfrac{3}{4}\,\text{inches}}{150\,\text{miles}} = \dfrac{4\dfrac{1}{2}\,\text{inches}}{x\,\text{miles}}$

$$\frac{3}{4}\,x = 4\frac{1}{2} \cdot 150$$

$$\frac{3}{4}\,x = 675$$

$$\frac{\dfrac{\cancel{3}^{1}}{\cancel{4}}\,x}{\dfrac{\cancel{3}^{1}}{\cancel{4}}} = \frac{675}{\dfrac{3}{4}}$$

$$x = 900\,\text{miles}$$

2. $\dfrac{2.4\,\text{cups}}{6\,\text{people}} = \dfrac{x\,\text{cups}}{2\,\text{people}}$

$$6x = 2 \cdot 2.4$$

$$6x = 4.8$$

$$\frac{\cancel{6}^{1}\,x}{\cancel{6}^{1}} = \frac{4.8}{6}$$

$$x = 0.8\,\text{cups}$$

3. $\dfrac{12\,\text{shirts}}{\$280} = \dfrac{15\,\text{shirts}}{x}$

$$12x = 15 \cdot 280$$

$$12x = 4{,}200$$

$$\frac{\cancel{12}^{1}\,x}{\cancel{12}^{1}} = \frac{4{,}200}{12}$$

$$x = \$350$$

4. $\dfrac{3 \text{ pounds}}{1{,}320 \text{ sq ft}} = \dfrac{x \text{ pounds}}{3{,}080 \text{ sq ft}}$

$$1{,}320x = 3 \cdot 3{,}080$$

$$1{,}320x = 9{,}240$$

$$\dfrac{\cancel{1{,}320}^{1} x}{\cancel{1{,}320}^{1}} = \dfrac{9{,}240}{1{,}320}$$

$$x = 7 \text{ pounds}$$

5. $\dfrac{8{,}100 \text{ miles}}{9 \text{ months}} = \dfrac{x}{15 \text{ months}}$

$$9x = 8{,}100 \cdot 15$$

$$9x = 121{,}500$$

$$\dfrac{\cancel{9}^{1} x}{\cancel{9}^{1}} = \dfrac{121{,}500}{9}$$

$$x = 13{,}500 \text{ miles}$$

6. $\dfrac{2 \text{ gallons}}{500 \text{ sq ft}} = \dfrac{x \text{ gallons}}{1{,}460 \text{ sq ft}}$

$$500x = 2 \cdot 1{,}460$$

$$500x = 2{,}920$$

$$\dfrac{\cancel{500}^{1} x}{\cancel{500}^{1}} = \dfrac{2{,}920}{500}$$

$$x = 5.84 \text{ gallons (rounded to 6 gallons since you need to purchase whole gallons)}$$

7. $\dfrac{216 \text{ miles}}{9 \text{ days}} = \dfrac{x \text{ miles}}{14 \text{ days}}$

$$9x = 216 \cdot 14$$

$$9x = 3{,}024$$

$$\dfrac{\cancel{9}^{1} x}{\cancel{9}^{1}} = \dfrac{3{,}024}{9}$$

$$x = 336 \text{ miles}$$

8. $\dfrac{2 \text{ chapters}}{11 \text{ days}} = \dfrac{16 \text{ chapters}}{x \text{ days}}$

$$2x = 16 \cdot 11$$

$$2x = 176$$

$$\dfrac{\cancel{2}^{1}x}{\cancel{2}^{1}} = \dfrac{176}{2}$$

$$x = 88 \text{ days}$$

9. $\dfrac{\$456}{3 \text{ months}} = \dfrac{\$3{,}648}{x \text{ months}}$

$$456x = 3{,}648 \cdot 3$$

$$456x = 10{,}944$$

$$\dfrac{\cancel{456}^{1}x}{\cancel{456}^{1}} = \dfrac{10{,}944}{456}$$

$$x = 24 \text{ months}$$

10. $\dfrac{6 \text{ tickets}}{\$112.50} = \dfrac{x \text{ tickets}}{\$168.75}$

$$112.50x = 6 \cdot 168.75$$

$$112.50x = 1{,}012.50$$

$$\dfrac{\cancel{112.50}^{1}x}{\cancel{112.50}^{1}} = \dfrac{1{,}012.50}{112.50}$$

$$x = 9 \text{ tickets}$$

This section explained how to solve word problems using proportions. The most important part is setting up the proportion. Make sure that you have the same units in the numerators of both ratio statements and the same units in the denominators of the ratio statements.

Formulas

In mathematics and science, many problems can be solved by using a formula. A *formula* is a mathematical statement of the relationship of two or more variables. For example, the distance (D) an automobile travels is related to the rate (R) of speed and the time (T) it travels. In symbols, $D = RT$. To solve a word problem using a formula, simply select the correct formula, substitute the values of the variables, and evaluate the formula.

In order to evaluate formulas, you use the order of operations.

Step 1 Perform all operations in parentheses

Step 2 Raise each number to its power

Step 3 Perform multiplication and division from left to right

Step 4 Perform addition and subtraction from left to right

EXAMPLE

Find the interest on a loan whose principal (*P*) is $5,400 at a rate of 6% for eight years. Use *I = PRT*.

SOLUTION

Goal: You are being asked to find the interest.

Strategy: Use the formula *I = PRT*.

Implementation:

$$I = PRT$$
$$= \$5{,}400 \cdot 6\% \cdot 8$$
$$= \$5{,}400 \cdot (0.06) \cdot 8$$
$$= \$2{,}592$$

Evaluation: You can estimate the answer by rounding $5,400 to $5,000 and then finding 6% of $5,000, which is 0.06 × $5,000 = $300. The interest for one year is about $300. The interest for eight years then is 8 × $300 = $2,400. Since this is close to $2,592, the answer is probably correct.

EXAMPLE

Find the Fahrenheit temperature (F) when the Celsius temperature (C) is 90°. Use $F = \dfrac{9}{5}C + 32°$.

SOLUTION

Goal: You are being asked to find a Fahrenheit temperature.

Strategy: Use the formula $F = \dfrac{9}{5}C + 32°$.

Implementation:

$$F = \frac{9}{5}C + 32°$$

$$F = \frac{9}{5} \cdot \frac{90}{1} + 32°$$

$$= \frac{9}{\cancel{5}^1} \cdot \frac{\cancel{90}^{18}}{1} + 32°$$

$$= 9 \cdot 18 + 32°$$

$$= 194°$$

Evaluation: **You can estimate the answer by multiplying 90° by 1.5 and adding 30°. That is, 90 · 1.5 + 30 = 135 + 30 = 185°. Since 185° is close to 194°, your answer is probably correct.**

Still Struggling

Remember the order of operations. Always perform multiplication before addition.

EXAMPLE

The distance (*d*) an object falls in feet is $d = 32t^2$ where *t* is the time in seconds. Find the distance an object falls in five seconds.

SOLUTION

Goal: **You are being asked to find the distance an object falls in five seconds.**

Strategy: **Use the formula $d = 32t^2$.**

Implementation:

$$d = 32t^2$$

$$= 32(5)^2$$

$$= 32 \cdot 25$$

$$= 800 \text{ feet}$$

Evaluation: Estimate the answer by rounding 32 to 30 and then multiplying $30 \times 25 = 750$. Since this estimate is close to 800, the answer is probably correct.

Still Struggling

Remember the order of operations. Square before multiplying.

TRY THESE

1. Find the perimeter (P) of a square whose side (s) is 24 inches. Use $P = 4s$.

2. Find the current (I) in amperes when the electromotive force (E) is 15 volts and the resistance (R) is 9 ohms. Use $I = \dfrac{E}{R}$.

3. Find the volume (V) of a cylinder in cubic feet when the height (h) is 15 feet and the radius (r) of the base is 4 feet. Use $V = 3.14r^2h$.

4. Find the Celsius temperature (C) when the Fahrenheit temperature (F) is 59°. Use $C = \dfrac{5}{9}(F - 32°)$.

5. Find the force (F) of the wind against a flat surface whose area (A) is 32 square feet when the wind speed (s) is 30 miles per hour. Use $F = 0.004As^2$.

6. Find the Fahrenheit temperature (F) when the Celsius temperature (C) is 30°. Use $F = \dfrac{9}{5}C + 32°$.

7. Find the surface area (A) of a cube in square feet when each side (s) measures eight inches. Use $A = 6s^2$.

8. Find the amount of work (W) done by applying a force (F) of 80 pounds moving a distance (d) of 12 feet. Use $W = Fd$.

9. Find the distance (D) an automobile travels at a rate (R) of 42 miles per hour in 3.2 hours (T). Use $D = RT$.

10. Find the amount of interest (I) earned on a principal (P) of $8,220 at a rate ($R$) of 9% for a time ($T$) of four years. Use $I = PRT$.

✔ SOLUTIONS

1. $P = 4s$

    $= 4 \cdot 24$

    $= 96 \text{ inches}$

2. $I = \dfrac{E}{R}$

    $= \dfrac{15}{9}$

    $= 1\dfrac{2}{3} \text{ amperes}$

3. $V = 3.14 r^2 h$

    $= 3.14 \cdot 4^2 \cdot 15$

    $= 753.6 \text{ ft}^3$

4. $C = \dfrac{5}{9}(F - 32°)$

    $= \dfrac{5}{9}(59 - 32)$

    $= \dfrac{5}{9} \cdot 27$

    $= 15°$

5. $F = 0.004\, As^2$

    $= 0.004 \cdot 32 \cdot 30^2$

    $= 0.004 \cdot 32 \cdot 900$

    $= 115.2 \text{ pounds}$

6. $F = \dfrac{9}{5}C + 32°$

    $= \dfrac{9}{5}(30) + 32$

    $= 9 \cdot 6 + 32$

    $= 54 + 32$

    $= 86°$

7. $A = 6s^2$

    $= 6 \cdot 8^2$

    $= 6 \cdot 64$

    $= 384 \text{ sq ft}$

8. $W = Fd$

   $= 80 \cdot 12$

   $= 960$ ft-lbs

9. $D = RT$

   $= 42 \cdot 3.2$

   $= 134.4$ miles

10. $I = PRT$

    $= \$8{,}220 \cdot 9\% \cdot 4$

    $= \$8{,}220 \cdot 0.09 \cdot 4$

    $= \$2{,}959.20$

In this section, word problems were solved by using formulas. In order to use formulas correctly, you must follow the order of operations to evaluate formulas.

1. Perform all operations in parentheses.

2. Raise each number to a power.

3. Perform multiplication and division left to right.

4. Perform addition and subtraction left to right.

Summary

This chapter explained how to solve word problems using proportions and formulas. These types of problems occur in physics, chemistry, and life sciences courses as well as in business mathematics courses and other areas.

QUIZ

1. If three ounces of a cereal contain 210 calories, how many calories would be contained in eight ounces?

   A. 630
   B. 420
   C. 560
   D. 1,680

2. If a person can swim 5 laps in a pool in 3 minutes, how many laps can the person swim in 15 minutes?

   A. 30
   B. 18
   C. 10
   D. 25

3. If four bottles of water cost $5.20, how much will 12 bottles cost?

   A. $15.60
   B. $20.80
   C. $10.40
   D. $62.40

4. If three items cost $25, how many items could you buy for $125?

   A. 9
   B. 15
   C. 12
   D. 18

5. If a 6-foot pole casts a shadow of 3.5 feet, how tall is a tree that casts a shadow of 14 feet?

   A. 18 feet
   B. 12 feet
   C. 20 feet
   D. 24 feet

6. How far (in feet) will an object fall in six seconds? Use $d = 32t^2$ where t = the time in seconds.

   A. 192 feet
   B. 384 feet
   C. 1,152 feet
   D. 576 feet

7. If a person travels a distance of 540 miles at 30 miles per hour, find the time it will take the person to get there. Use $T = \dfrac{D}{R}$ where D = distance and R = rate.

   A. 27 hours

   B. 18 hours

   C. 14 hours

   D. 9 hours

8. The formula for finding the volume of a cylinder is $V = \pi r^2 h$ where π = 3.14, r = the value of the radius, and h = the height. What is the volume of a cylinder whose radius is three feet and whose height is four feet?

   A. 527.52 feet

   B. 37.68 feet

   C. 113.04 feet

   D. 75.36 feet

9. The interest on a loan is found by using the formula $I = PRT$ where P = the principle, R = the rate, and T = the time in years. How much interest would a person have to pay on a $3,250 loan at 5% for three years?

   A. $487.50

   B. $1562.50

   C. $975.00

   D. $643.00

10. Find the distance around a circular swimming pool if its diameter is seven feet. Use $C = \pi D$ where π = 3.14.

    A. 21 feet

    B. 10.99 feet

    C. 43.96 feet

    D. 21.98 feet

5

Equations and Algebraic Representation

This chapter reviews how to solve equations. Many of the word problems in algebra can be solved by setting up an equation based on the words in the problem and solving it. In order to set up an equation, you need to translate the information given in the problem into symbols. This is called algebraic representation. These two skills are very important when solving word problems.

CHAPTER OBJECTIVES

In this chapter, you will learn how to

- Solve equations
- Represent word statements using letters (variables) and mathematical symbols

Refresher IV: Equations

An *algebraic expression* consists of variables (letters), numbers, operation signs (+, −, ×, ÷), and grouping symbols. Here are a few examples of algebraic expressions:

$$3x \qquad 5(x - 6) \qquad -8x^2 \qquad 9 + 2$$

An *equation* is a statement of equality of two algebraic expressions. Here are some examples of equations:

$$5 + 4 = 9 \qquad 3x - 2 = 13 \qquad x^2 + 3x + 2 = 0$$

An equation that contains a variable is called a *conditional* equation. To *solve* a conditional equation, it is necessary to find a number that, when substituted for the variable, makes a true equation. This number is called a *solution* to the equation. For example, 5 is a solution to the equation $x + 3 = 8$ since when 5 is substituted for x, it makes the equation true; that is, $5 + 3 = 8$. The process of finding a solution to an equation is called *solving* the equation. To *check* an equation, substitute the solution into the original equation and see if it's a true equation.

There are four types of basic equations. In order to solve each type, you perform the opposite operation to both sides of the equation as the operation that is being performed on the variable. Addition and subtraction are opposite operations. Multiplication and division are opposite operations. The next four examples show how to solve basic equations.

EXAMPLE

Solve for x: $x - 10 = 13$

SOLUTION

$$x - 10 = 13$$
$$x - 10 + 10 = 13 + 10 \qquad \text{Add 10 to both sides}$$
$$x = 23$$

Check: $x - 10 = 13$
$$23 - 10 = 13$$
$$13 = 13$$

EXAMPLE

Solve for x: $x + 16 = 34$

SOLUTION

$$x + 16 = 34$$

$$x + 16 - 16 = 34 - 16 \qquad \text{Subtract 16 from both sides}$$

$$x = 18$$

Check: $x + 16 = 34$

$$18 + 16 = 34$$

$$34 = 34$$

EXAMPLE

Solve for x: $\dfrac{x}{7} = 4$

SOLUTION

$$\frac{x}{7} = 4$$

$$\frac{7^1}{1} \cdot \frac{x}{7^1} = 4 \cdot 7 \qquad \text{Multiply both sides by 7}$$

$$x = 28$$

Check: $\dfrac{x}{7} = 4$

$$\frac{28}{7} = 4$$

$$4 = 4$$

EXAMPLE

Solve for x: $8x = 56$

SOLUTION

$$8x = 56$$

$$\frac{8^1 x}{8^1} = \frac{56}{8}$$

$$x = 7$$

Check: $8x = 56$

$8 \cdot 7 = 56$

$56 = 56$

? Still Struggling

When solving equations, the same number (except zero) can be added to, subtracted from, multiplied by, or divided into both sides of the equation without changing the equality of the equation.

TRY THESE

Solve each equation for x:

1. $x + 32 = 56$

2. $x - 12 = 7$

3. $6x = 72$

4. $x + 4 = 16$

5. $\dfrac{x}{5} = 19$

SOLUTIONS

1. $\qquad x + 32 = 56$

   $x + 32 - 32 = 56 - 32$

   $\qquad\qquad x = 24$

2. $\qquad x - 12 = 7$

   $x - 12 + 12 = 7 + 12$

   $\qquad\qquad x = 19$

3. $\qquad 6x = 72$

   $\dfrac{\cancel{6}^{1} x}{\cancel{6}^{1}} = \dfrac{72}{6}$

   $\qquad x = 12$

4. $x + 4 = 16$

$x + 4 - 4 = 16 - 4$

$x = 12$

5. $\dfrac{x}{5} = 19$

$\dfrac{\cancel{5}^{1}}{1} \cdot \dfrac{x}{\cancel{5}^{1}} = 19 \cdot 5$

$x = 95$

More complex equations require several steps to solve. The goal is to use addition and/or subtraction to get the variables on one side of the equation and the numbers on the other side of the equation. Then divide both sides by the number in front of the variable. This number is called the numerical coefficient of the variable.

EXAMPLE

Solve for x: $3x + 18 = 42$

SOLUTION

$3x + 18 = 42$

$3x + 18 - 18 = 42 - 18$ Subtract 18

$3x = 24$

$\dfrac{\cancel{3}^{1}x}{\cancel{3}^{1}} = \dfrac{24}{3}$ Divide by 3

$x = 8$

Check: $3x + 18 = 42$

$3(8) + 18 = 42$

$24 + 18 = 42$

$42 = 42$

EXAMPLE

Solve for x: $9x - 15 = 4x + 35$

SOLUTION

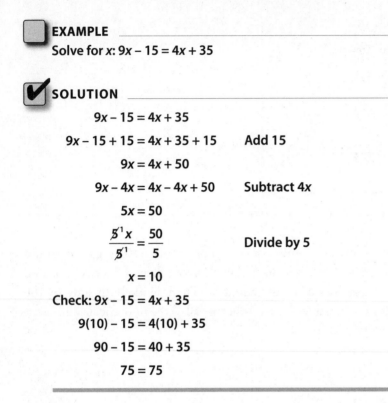

$$9x - 15 = 4x + 35$$

$$9x - 15 + 15 = 4x + 35 + 15 \qquad \text{Add 15}$$

$$9x = 4x + 50$$

$$9x - 4x = 4x - 4x + 50 \qquad \text{Subtract } 4x$$

$$5x = 50$$

$$\frac{\cancel{5}^{1} x}{\cancel{5}^{1}} = \frac{50}{5} \qquad \text{Divide by 5}$$

$$x = 10$$

Check: $9x - 15 = 4x + 35$

$$9(10) - 15 = 4(10) + 35$$

$$90 - 15 = 40 + 35$$

$$75 = 75$$

? Still Struggling

If you get the wrong answer, it is generally better to start over and solve the equation again rather than trying to find your mistake. This is especially true when there are several steps in the solution.

TRY THESE

Solve each equation for x:

1. $12x + 21 = 81$

2. $4x - 10 = 14$

3. $15x - 21 = 12x + 18$

4. $7x + 6 = 3x + 48$

5. $11x - 9 = 5x + 33$

✔ SOLUTIONS

1. $12x + 21 = 81$

    $12x + 21 - 21 = 81 - 21$

    $12x = 60$

    $$\frac{\cancel{12}^{1}x}{\cancel{12}^{1}} = \frac{60}{12}$$

    $x = 5$

2. $4x - 10 = 14$

    $4x - 10 + 10 = 14 + 10$

    $4x = 24$

    $$\frac{\cancel{4}^{1}x}{\cancel{4}^{1}} = \frac{24}{4}$$

    $x = 6$

3. $15x - 21 = 12x + 18$

    $15x - 21 + 21 = 12x + 18 + 21$

    $15x = 12x + 39$

    $15x - 12x = 12x - 12x + 39$

    $3x = 39$

    $$\frac{\cancel{3}^{1}x}{\cancel{3}^{1}} = \frac{39}{3}$$

    $x = 13$

4. $7x + 6 = 3x + 48$

    $7x + 6 - 6 = 3x + 48 - 6$

    $7x = 3x + 42$

    $7x - 3x = 3x - 3x + 42$

    $4x = 42$

    $$\frac{\cancel{4}^{1}x}{\cancel{4}^{1}} = \frac{42}{4}$$

    $x = 10.5$

5. $11x - 9 = 5x + 33$

$11x - 9 + 9 = 5x + 33 + 9$

$11x = 5x + 42$

$11x - 5x = 5x - 5x + 42$

$6x = 42$

$$\frac{\cancel{6}^1 x}{\cancel{6}^1} = \frac{42}{6}$$

$x = 7$

Many equations contain parentheses. In order to remove parentheses, multiply each term inside the parentheses by the number outside the parentheses. This is called the distributive property of multiplication over addition. For example,

$$4(x + 8) = 4 \cdot x + 4 \cdot 8 = 4x + 32$$

$$9(3x - 7) = 9 \cdot 3x - 9 \cdot 7 = 27x - 63$$

When you solve an equation, remove parentheses first, combine like terms (i.e., $6x + 8x$), and then solve as shown in the previous examples. Like terms have the same variables that are raised to the same powers.

EXAMPLE

Solve for x: $8(3x - 4) = 88$

SOLUTION

$8(3x - 4) = 88$

$8 \cdot 3x - 8 \cdot 4 = 88$ **Remove parentheses**

$24x - 32 = 88$

$24x - 32 + 32 = 88 + 32$ **Add 32**

$24x = 120$

$$\frac{\cancel{24}^1 x}{\cancel{24}^1} = \frac{120}{24}$$ **Divide by 24**

$x = 5$

Check: $8(3x - 4) = 88$

$8(3 \cdot 5 - 4) = 88$

$8(15 - 4) = 88$

$8(11) = 88$

$88 = 88$

EXAMPLE

Solve for x: $6(2x + 7) - 10x = 56$

SOLUTION

$6(2x + 7) - 10x = 56$

$6 \cdot 2x + 6 \cdot 7 - 10x = 56$

$12x + 42 - 10x = 56$ Remove parentheses

$2x + 42 = 56$ Combine $12x - 10x$

$2x + 42 - 42 = 56 - 42$ Subtract 42

$2x = 14$

$$\frac{\cancel{2}^{1}x}{\cancel{2}^{1}} = \frac{14}{2}$$ Divide by 2

$x = 7$

Check: $6(2x + 7) - 10x = 56$

$6(2 \cdot 7 + 7) - 10(7) = 56$

$6(14 + 7) - 70 = 56$

$6(21) - 70 = 56$

$126 - 70 = 56$

$56 = 56$

TRY THESE

Solve each equation for x:

1. $4(x - 10) + 20 = 28$

2. $3(3x - 7) = 7x + 33$

3. $9(4x - 5) = 99$

4. $8(2x - 7) = 7(2x + 4)$

5. $2(4x + 9) = 5x - 3$

✔ SOLUTIONS

1. $4(x - 10) + 20 = 28$

     $4 \cdot x - 4 \cdot 10 + 20 = 28$

     $4x - 40 + 20 = 28$

     $4x - 20 = 28$

     $4x - 20 + 20 = 28 + 20$

     $4x = 48$

$$\frac{\cancel{4}^{1}x}{\cancel{4}^{1}} = \frac{48}{4}$$

     $x = 12$

2. $3(3x - 7) = 7x + 33$

     $3 \cdot 3x - 3 \cdot 7 = 7x + 33$

     $9x - 21 = 7x + 33$

     $9x - 7x - 21 = 7x - 7x + 33$

     $2x - 21 = 33$

     $2x - 21 + 21 = 33 + 21$

     $2x = 54$

$$\frac{\cancel{2}^{1}x}{\cancel{2}^{1}} = \frac{54}{2}$$

     $x = 27$

3. $9(4x - 5) = 99$

     $9 \cdot 4x - 9 \cdot 5 = 99$

     $36x - 45 = 99$

     $36x - 45 + 45 = 99 + 45$

     $36x = 144$

$$\frac{\cancel{36}^{1}x}{\cancel{36}^{1}} = \frac{144}{36}$$

     $x = 4$

4. $8(2x - 7) = 7(2x + 4)$

$8 \cdot 2x - 8 \cdot 7 = 7 \cdot 2x + 7 \cdot 4$

$16x - 56 = 14x + 28$

$16x - 14x - 56 = 14x - 14x + 28$

$2x - 56 = 28$

$2x - 56 + 56 = 28 + 56$

$2x = 84$

$\dfrac{\cancel{2}^{1}x}{\cancel{2}^{1}} = \dfrac{84}{2}$

$x = 42$

5. $2(4x + 9) = 5x - 3$

$2 \cdot 4x + 2 \cdot 9 = 5x - 3$

$8x + 18 = 5x - 3$

$8x - 5x + 18 = 5x - 5x - 3$

$3x + 18 = -3$

$3x + 18 - 18 = -3 - 18$

$3x = -21$

$\dfrac{\cancel{3}^{1}x}{\cancel{3}^{1}} = \dfrac{-21}{3}$

$x = -7$

Sometimes when you are solving word problems, you will need to solve an equation containing fractions. It should be noted that fraction terms can be written in two ways. See the next examples:

$\dfrac{1}{4}x$ can be written as $\dfrac{x}{4}$

$\dfrac{3}{5}x$ can be written as $\dfrac{3x}{5}$

$\dfrac{5}{6}(x - 3)$ can be written as $\dfrac{5(x - 3)}{6}$

To solve an equation containing fractions, it is necessary to find the lowest common denominator of all the fractions, and then multiply each term in the equation by the lowest common denominator. This process is called *clearing fractions*.

EXAMPLE

Solve for x: $\dfrac{x}{4} + \dfrac{x}{6} = 25$

SOLUTION

$$\dfrac{x}{4} + \dfrac{x}{6} = 25 \qquad \text{The LCD is 12}$$

$$\dfrac{\overset{3}{\cancel{12}}}{1} \cdot \dfrac{x}{\underset{1}{\cancel{4}}} + \dfrac{\overset{2}{\cancel{12}}}{1} \cdot \dfrac{x}{\underset{1}{\cancel{6}}} = 25 \cdot 12 \qquad \text{Clear fractions}$$

$$3x + 2x = 300$$

$$5x = 300 \qquad \text{Combine like terms}$$

$$\dfrac{\overset{1}{\cancel{5}}x}{\underset{1}{\cancel{5}}} = \dfrac{300}{5} \qquad \text{Divide by 5}$$

$$x = 60$$

$$\text{Check: } \dfrac{x}{4} + \dfrac{x}{6} = 25$$

$$\dfrac{60}{4} + \dfrac{60}{6} = 25$$

$$15 + 10 = 25$$

$$25 = 25$$

EXAMPLE

Solve for x: $\dfrac{1}{5} + \dfrac{2}{3} = \dfrac{1}{x}$

SOLUTION

$$\dfrac{1}{5} + \dfrac{2}{3} = \dfrac{1}{x} \qquad \text{The LCD is } 15x$$

$$\dfrac{\overset{3}{\cancel{15}}x}{1} \cdot \dfrac{1}{\underset{1}{\cancel{5}}} + \dfrac{\overset{5}{\cancel{15}}x}{1} \cdot \dfrac{2}{\underset{1}{\cancel{3}}} = \dfrac{15x^{\overset{1}{}}}{1} \cdot \dfrac{1}{x^{\underset{1}{}}} \qquad \text{Clear fractions}$$

$$3x + 10x = 15 \qquad \text{Combine terms}$$

$$13x = 15$$

$$\dfrac{\overset{1}{\cancel{13}}x}{\underset{1}{\cancel{13}}} = \dfrac{15}{13} \qquad \text{Divide by 13}$$

$$x = \dfrac{15}{13} = 1\dfrac{2}{13}$$

Check: $\dfrac{1}{5} + \dfrac{2}{3} = \dfrac{1}{x}$

$$\frac{1}{5} + \frac{2}{3} = \frac{1}{1\frac{2}{13}}$$

$$\frac{1}{5} + \frac{2}{3} = \frac{13}{15}$$

$$\frac{3}{15} + \frac{10}{15} = \frac{13}{15}$$

$$\frac{13}{15} = \frac{13}{15}$$

TRY THESE

Solve each equation for x:

1. $\dfrac{x}{3} + \dfrac{x}{5} + \dfrac{x}{2} = 93$

2. $\dfrac{3}{4}x - \dfrac{2}{3} = \dfrac{5}{6}$

3. $\dfrac{1}{3} + \dfrac{1}{8} = \dfrac{1}{x}$

4. $\dfrac{1}{4}(2x - 6) = \dfrac{3}{8}$

5. $\dfrac{1}{5}x + 20 = \dfrac{1}{2}x$

SOLUTIONS

1. $$\frac{x}{3} + \frac{x}{5} + \frac{x}{2} = 93$$

$$\frac{\overset{10}{\cancel{30}}}{1} \cdot \frac{x}{\cancel{3}^1} + \frac{\overset{6}{\cancel{30}}}{1} \cdot \frac{x}{\cancel{5}^1} + \frac{\overset{15}{\cancel{30}}}{1} \cdot \frac{x}{\cancel{2}^1} = 93 \cdot 30$$

$$10x + 6x + 15x = 2790$$

$$31x = 2790$$

$$\frac{\cancel{31}^{\,1} x}{\cancel{31}_{\,1}} = \frac{2790}{31}$$

$$x = 90$$

2.
$$\frac{3}{4}x - \frac{2}{3} = \frac{5}{6}$$

$$\frac{\cancel{12}^{\,3}}{1} \cdot \frac{3}{\cancel{4}^{\,1}}x - \frac{\cancel{12}^{\,4}}{1} \cdot \frac{2}{\cancel{3}^{\,1}} = \frac{\cancel{12}^{\,2}}{1} \cdot \frac{5}{\cancel{6}^{\,1}}$$

$$9x - 8 = 10$$

$$9x - 8 + 8 = 10 + 8$$

$$9x = 18$$

$$\frac{\cancel{9}^{\,1} x}{\cancel{9}^{\,1}} = \frac{18}{9}$$

$$x = 2$$

3.
$$\frac{1}{3} + \frac{1}{8} = \frac{1}{x}$$

$$\frac{\cancel{24}^{\,8} x}{1} \cdot \frac{1}{\cancel{3}^{\,1}} + \frac{\cancel{24}^{\,3} x}{1} \cdot \frac{1}{\cancel{8}^{\,1}} = 24\cancel{x}^{\,1} \cdot \frac{1}{\cancel{x}^{\,1}}$$

$$8x + 3x = 24$$

$$11x = 24$$

$$\frac{\cancel{11}^{\,1} x}{\cancel{11}^{\,1}} = \frac{24}{11}$$

$$x = \frac{24}{11} = 2\frac{2}{11}$$

4.
$$\frac{1}{4}(2x - 6) = \frac{3}{8}$$

$$\frac{\cancel{8}^{\,2}}{1} \cdot \frac{1}{\cancel{4}^{\,1}}(2x - 6) = \frac{\cancel{8}^{\,1}}{1} \cdot \frac{3}{\cancel{8}^{\,1}}$$

$$2(2x - 6) = 3$$

$$4x - 12 = 3$$

$$4x - 12 + 12 = 3 + 12$$

$$4x = 15$$

$$\frac{\cancel{4}^1 x}{\cancel{4}^1} = \frac{15}{4}$$

$$x = 3\frac{3}{4}$$

5. $$\frac{1}{5}x + 20 = \frac{1}{2}x$$

$$\frac{\cancel{10}^2}{1} \cdot \frac{1x}{\cancel{5}^1} + 10 \cdot 20 = \frac{\cancel{10}^5}{1} \cdot \frac{1}{\cancel{2}^1}x$$

$$2x + 200 = 5x$$

$$2x - 2x + 200 = 5x - 2x$$

$$200 = 3x$$

$$\frac{200}{3} = \frac{\cancel{3}^1 x}{\cancel{3}^1}$$

$$\frac{200}{3} = x$$

$$66\frac{2}{3} = x$$

Algebraic Representation

When you solve an algebra word problem, you must first be able to translate the conditions of the problem into an equation involving algebraic expressions. Recall that an algebraic expression will consist of variables (letters), numbers, operations signs $(+, -, \times, \div)$, and grouping symbols such as parentheses.

Here are some common phrases that are used in algebra word problems:

Addition can be denoted by
sum
added to

increased by
larger than
more than

Subtraction can be denoted by
less than
subtracted from
decreased by
exceeds
shorter than
difference between

Multiplication can be denoted by
product
times
multiplied by
twice as large
three times a number
1/2 of a number

Division can be denoted by
divided by
quotient of

Equals can be denoted by
is
will be
is equal to

Here are some examples of how these phrases are translated into symbols:

Word Statement	*Symbolic Representation*
A number increased by 8	$x + 8$
Six times a number	$6x$
Seven less than a number	$x - 7$
One-third of a number	$\frac{1}{3}x$ or $\frac{x}{3}$
The square of a number	x^2
Four times a number increased by 9	$4x + 9$

Six times a number less 2 $6x - 2$

The cost of y yards of hose at \$0.39 a yard $0.39\,y$

TRY THESE

Write each in symbols:

1. A number increased by 10

2. Four times a number plus 8

3. Six less than a number

4. Five less than six times a number

5. The square of a number decreased by 4

SOLUTIONS

1. $x + 10$

2. $4x + 8$

3. $x - 6$

4. $6x - 5$

5. $x^2 - 4$

In the previous examples, only one unknown was used. Other times, it is necessary to represent two *related* unknowns by using one variable. Consider these examples:

"The sum of two numbers is 15." When you are given two numbers whose sum is 15 and one number is, say, 9, how would you find the other number? You would find $15 - 9$. So if one number is x, the other number would be $15 - x$.

"One number is 5 more than another number." If I told you one number is 12, how would you find the other number? You would add $12 + 5$. So if one stated number is x, the other number would be $x + 5$.

"One number is two times another number." If I told you one number is 4, how would you find the other number? You would multiply 4 by 2. So if one number is x, the other number would be $2x$.

"One number is 7 less than another number." If I told you one number was 22, how would you find the other number? You would subtract $22 - 7$. So if one number is x, the other number is $x - 7$.

? Still Struggling

When you are representing two related numbers, you have choices on how you do it. For example, if the problem says represent two numbers such that one number is 7 more than the other number, you could represent them as x and $x + 7$ or x and $x - 7$. Either way is correct.

TRY THESE

Represent each using symbols:

1. The sum of two numbers is 24.

2. One number is 6 less than the other number.

3. The second number is 5 less than one-third of the first number.

4. The second number is 8 more than twice the first number.

5. The second number is three times the first number.

SOLUTIONS

1. Let x = the first number and $24 - x$ = the second number.

2. The first number is x, and the second number is $x - 6$.

3. Let x = the first number and $\frac{1}{3}x - 5$ = the second number.

4. Let x = the first number and $2x + 8$ = the second number.

5. Let x = the first number and $3x$ = the second number.

Now that you know how to translate word phrases into algebraic expressions, the next step is to translate whole sentences into algebraic expressions using the equal sign. Consider these examples.

"Five times a number increased by 8 is equal to 38" translates to

$$5x + 8 = 38$$

"Nine times a number decreased by 4 is equal to 32."

$$9x - 4 = 32$$

"The difference between a number and one-fourth itself is equal to 20."

$$x - \frac{1}{4}x = 20$$

TRY THESE

Translate each into an equation:

1. **Three times a number decreased by 7 is 17.**

2. **If 4 is added to a number, you get 16.**

3. **The sum of a number and three times itself is equal to 32.**

4. **One-fourth a number plus 6 is 54.**

5. **If 8 is increased by twice a number, the sum is 26.**

SOLUTIONS

1. $3x - 7 = 17$

2. $x + 4 = 16$

3. $x + 3x = 32$

4. $\frac{1}{4}x + 6 = 54$

5. $8 + 2x = 26$

Finally, it is necessary to be able to write an equation for two related unknowns using one variable.

EXAMPLE

Write an equation for this problem: "One number is 8 more than another number and their sum is 17."

SOLUTION

Let x = the smaller number and $x + 8$ = the larger number. The equation is $x + x + 8 = 17$.

EXAMPLE

Write an equation for this problem: "One number is four times as large as another number. If two times the smaller number is subtracted from the larger number, the answer is 18."

 SOLUTION

Let x = the smaller number and $4x$ = the larger number. The equation is $4x - 2x = 18$.

 TRY THESE

Write an equation for each. Do not solve the equations.

1. One number is 6 more than three times another number. Find the numbers if their sum is 66.

2. One number is $\dfrac{1}{2}$ of another number. Find the numbers if their sum is 36.

3. What number increased by $\dfrac{1}{3}$ of itself is equal to 9?

4. Two times a number is 9 more than $\dfrac{1}{2}$ the number. Find the numbers.

5. A certain number exceeds another number by 10. If their sum is 63, find the numbers.

 SOLUTIONS

1. Let x = one number and $3x + 6$ = the other number. The equation is $x + 3x + 6 = 66$.

2. Let x = one number and $\dfrac{1}{2}x$ = the other number. The equation is $x + \dfrac{1}{2}x = 36$.

3. Let x = the number and $\dfrac{1}{3}x$ = the other number. The equation is $x + \dfrac{1}{3}x = 9$.

4. Let x = the number. The equation is $2x = \dfrac{1}{2}x + 9$.

5. Let x = one number and $x + 10$ = the larger number. The equation is $x + x + 10 = 63$.

Summary

The first part of this chapter explained how to solve equations. Many types of word problems are solved using equations. It is a very important algebraic topic.

This chapter also explains a very important skill that is used to solve word problems in mathematics. That skill is being able to translate the words of the problem into mathematical symbols and to set up an equation using these symbols. Once the equation is obtained, all that is necessary to get the answer is to algebraically solve the equation for the variable.

QUIZ

1. The solution to the equation $11x + 20 = 6x - 25$ is
   A. $x = 9$
   B. $x = -11$
   C. $x = -9$
   D. $x = 11$

2. The solution to the equation $6 - x = -14 - 3x$ is
   A. $x = 10$
   B. $x = 5$
   C. $x = -3$
   D. $x = -10$

3. The solution to the equation $x - 9 = 15 - 3x$ is
   A. $x = 6$
   B. $x = 3$
   C. $x = 8$
   D. $x = 2$

4. The solution to the equation $5(x - 6) + 3(2 - x) = 0$ is
   A. $x = 14$
   B. $x = 12$
   C. $x = -12$
   D. $x = 16$

5. The solution to the equation $9(7x - 3) = 31 + 8(7x - 2)$ is
   A. $x = 5$
   B. $x = 3$
   C. $x = 6$
   D. $x = 4$

6. The statement 3 times a number x plus 6 can be represented as
   A. $6x + 3$
   B. $9x$
   C. $3x + 6$
   D. $3x - 6$

7. The statement "12 less than five times a number x" can be represented as
   A. $12x - 5$
   B. $5x - 12$
   C. $5 \cdot 12x$
   D. $(5 + 12)x$

8. **If the sum of two numbers is 32, and one number is x, the other number is**
   A. $32 - x$
   B. $x - 32$
   C. $32x$
   D. $32 \div x$

9. **The equation represented by the statement "five times a number plus 8 is equal to 45" is**
   A. $5x - 8 = 45$
   B. $8 \cdot 5x = 45$
   C. $5(x + 8) = 45$
   D. $5x + 8 = 45$

10. **The equation represented by the statement "16 minus two times a number is equal to 24" is**
    A. $2 - 16x = 24$
    B. $16 - 2x = 24$
    C. $16x - 2 = 24$
    D. $2x - 16 = 24$

chapter 6

Solving Number and Digit Problems

This chapter explains how to solve number and digit problems. The numbers used are most often whole numbers and are usually positive. The problem will give you the relationship between two or more numbers in order to write an equation, and then solve the equation.

The numbers we use today use the digits 0 through 9. We have one-digit numbers, two-digit numbers, three-digit numbers, etc. Knowing what each digit is in a number will enable you to write an equation in order to solve the problem.

CHAPTER OBJECTIVES

In this chapter, you will learn how to

- Solve word problems about numbers
- Solve digit problems

Number Problems

The strategy used to solve word problems in algebra is as follows:

1. Represent an unknown by using a letter such as x.
2. If necessary, represent the other unknowns by using algebraic expressions in terms of x.
3. From the conditions of the problem, write an equation using the algebraic representation of the unknown(s).
4. Solve the equation for x.

EXAMPLE

One number is 6 less than another number and the sum of the two numbers is 32. Find the numbers.

SOLUTION

Goal: You are being asked to find two numbers.

Strategy: Let x = one number and $x - 6$ = the other number. Since the problem asked for the sum, write the equation as $x + x - 6 = 32$.

Implementation: Solve the equation for x:

$$x + x - 6 = 32$$
$$2x - 6 = 32$$
$$2x - 6 + 6 = 32 + 6$$
$$2x = 38$$
$$\frac{2^1 x}{2^1} = \frac{38}{2}$$
$$x = 19$$

Hence, one number is 19 and the other number is $x - 6$ or $19 - 6 = 13$.

Evaluation: Check the answer: $19 + 13 = 32$

EXAMPLE

If 5 plus three times a number is equal to 32, find the number.

SOLUTION

Goal: You are being asked to find one number.

Strategy: Let x = the number. Five plus three times a number is written as $5 + 3x$, and the equation is $5 + 3x = 32$.

Implementation: Solve the equation.

$$5 + 3x = 32$$
$$5 - 5 + 3x = 32 - 5$$
$$3x = 27$$
$$\frac{\cancel{3}^{1}x}{\cancel{3}^{1}} = \frac{27}{3}$$
$$x = 9$$

Evaluation: Check the answer: $5 + 3 \cdot 9 = 32$

EXAMPLE

A professor has two mathematics classes with a total of 46 students in both classes. If there are six more students in one class than the other, how many students are in each class?

SOLUTION

Goal: You are being asked to find the number of students in each of two classes.

Strategy: Let x = the number of students in one class and $x + 6$ be the number of students in the other class. The equation is $x + x + 6 = 46$.

Implementation: Solve the equation.

$$x + x + 6 = 46$$
$$2x + 6 = 46$$
$$2x + 6 - 6 = 46 - 6$$
$$2x = 40$$
$$\frac{\cancel{2}^{1}x}{\cancel{2}^{1}} = \frac{40}{2}$$
$$x = 20$$
$$x + 6 = 26$$

Hence, there are 20 students in one class and 26 students in the other.

Evaluation: Check the answer: $20 + 26 = 46$

EXAMPLE

If a number is decreased by 4 and two times the original number is equal to six times the other number, find the numbers.

SOLUTION

Goal: You are being asked to find two numbers.

Strategy: Let x = the original number and $x - 4$ = the other number.

Now two times the original number is $2x$ and six times the other number is $6(x - 4)$. The equation is $2x = 6(x - 4)$.

Implementation: Solve the equation.

$$2x = 6(x - 4)$$
$$2x = 6x - 24$$
$$2x - 6x = 6x - 6x - 24$$
$$-4x = -24$$
$$\frac{-\cancel{4}^1 x}{-\cancel{4}^1} = \frac{-24}{-4}$$
$$x = 6$$
$$x - 4 = 6 - 4 = 2$$

Hence, the first number is 6 and the second number is 2.

Evaluation: Check the answer: $6 - 4 = 2$ and $2 \cdot 6 = 6 \cdot 2$ or $12 = 12$.

Some number problems use *consecutive integers*. Numbers such as 1, 2, 3, 4, 5, etc., are called consecutive integers. They differ by 1. Consecutive integers can be represented as:

Let x = the first integer

$x + 1$ = the second integer

$x + 2$ = the third integer

etc.

Consecutive odd integers are numbers such as 1, 3, 5, 7, 9, 11, etc. They differ by 2. They can be represented as:

Let x = the first odd integer

$x + 2$ = the second consecutive odd integer

$x + 4$ = the third consecutive odd integer

etc.

Consecutive even integers are numbers such as 2, 4, 6, 8, 10, 12, etc. They also differ by 2. They can be represented as:

Let x = the first even integer

$x + 2$ = the second consecutive even integer

$x + 4$ = the third consecutive even integer

etc.

Still Struggling

You need not worry whether you are looking for consecutive even or odd numbers since the problems will always work out correctly. (The textbook authors have made them up so that they will.)

EXAMPLE

Find three consecutive integers whose sum is 96.

SOLUTION

Goal: You are being asked to find three consecutive integers whose sum is 96.

Strategy: Let x = the first integer, $x + 1$ = the second integer, and $x + 2$ = the third integer. The equation is $x + (x + 1) + (x + 2) = 96$.

Implementation: Solve the equation:

$$x + x + 1 + x + 2 = 96$$
$$3x + 3 = 96$$
$$3x + 3 - 3 = 96 - 3$$
$$3x = 93$$
$$x = 31$$
$$x + 1 = 31 + 1 = 32$$
$$x + 2 = 31 + 2 = 33$$

Evaluation: Check the answer: $31 + 32 + 33 = 96$

EXAMPLE

If the sum of two consecutive odd integers is 36, find the numbers.

SOLUTION

Goal: You are being asked to find two consecutive odd integers whose sum is 36.

Strategy: Let x = the first consecutive odd integer and $x + 2$ = the second consecutive odd integer. Since the sum is equal to 36, the equation is $x + x + 2 = 36$.

Implementation: Solve the equation:

$$x + x + 2 = 36$$
$$2x + 2 = 36$$
$$2x + 2 - 2 = 36 - 2$$
$$2x = 34$$
$$\frac{\cancel{2}^1 x}{\cancel{2}^1} = \frac{34}{2}$$
$$x = 17 \qquad \text{First odd integer}$$
$$x + 2 = 17 + 2 = 19 \qquad \text{Second odd integer}$$

Evaluation: 17 and 19 are consecutive odd integers, and their sum is $17 + 19 = 36$.

TRY THESE

1. A baseball team played 27 games and won five more games than it lost. Find the number of games the team won.

2. If three times a number plus 10 is equal to 22, find the number.

3. Four times a number decreased by 2 is equal to 26. Find the number.

4. If $\frac{1}{4}$ of a number is 12 less than $\frac{1}{3}$ of the number, find the number.

5. A carpenter wants to cut a 42-inch piece of lumber into three pieces so that each piece is six inches longer than the preceding one. Find the length of each piece.

6. The difference of two numbers is 45, and one number is six times the other number, find the numbers.

7. Find two numbers whose sum is 25 and whose difference is 3.

8. Fifty notebooks are placed into two boxes so that one box has six more notebooks than the other box. How many notebooks are in each box?

9. If the sum of three consecutive integers is 81, find the numbers.

10. The sum of two consecutive even integers is 62. Find the numbers.

✔ SOLUTIONS

1. Let x = the number of games the team lost and $x + 5$ = the number of games the team won.

$$x + x + 5 = 27$$
$$2x + 5 = 27$$
$$2x + 5 - 5 = 27 - 5$$
$$2x = 22$$
$$\frac{\cancel{2}^1 x}{\cancel{2}^1} = \frac{22}{2}$$
$$x = 11$$
$$x + 5 = 16$$

The team won 16 games and lost 11 games.

2. Let x = the number, then $3x + 10 = 22$.

$$3x + 10 = 22$$
$$3x + 10 - 10 = 22 - 10$$
$$3x = 12$$
$$\frac{\cancel{3}^1 x}{\cancel{3}^1} = \frac{12}{3}$$
$$x = 4$$

3. Let x = the number, then $4x - 2 = 26$

$$4x - 2 = 26$$
$$4x - 2 + 2 = 26 + 2$$
$$4x = 28$$
$$\frac{\cancel{4}^1 x}{\cancel{4}^1} = \frac{28}{4}$$
$$x = 7$$

4. Let x = the number

$$\frac{1}{3}x = \frac{1}{4}x + 12$$

$$\frac{\cancel{12}^{4}}{1} \cdot \frac{1x}{\cancel{3}^{1}} = \frac{\cancel{12}^{3}}{1} \cdot \frac{1}{\cancel{4}^{1}}x + 12 \cdot 12$$

$$4x = 3x + 144$$

$$4x - 3x = 3x - 3x + 144$$

$$x = 144$$

5. Let x = the length of the first piece of lumber, $x + 6$ = the length of the second piece, and $x + 12$ = the length of the third piece. Then $x + x + 6 + x + 12 = 42$ inches.

$$x + x + 6 + x + 12 = 42$$

$$3x + 18 = 42$$

$$3x + 18 - 18 = 42 - 18$$

$$3x = 24$$

$$\frac{\cancel{3}^{1}x}{\cancel{3}^{1}} = 24$$

$$x = 8 \text{ inches}$$

$$x + 6 = 14 \text{ inches}$$

$$x + 12 = 20 \text{ inches}$$

6. Let x = one number and $6x$ = the other number. Then $6x - x = 45$.

$$6x - x = 45$$

$$5x = 45$$

$$\frac{\cancel{5}^{1}x}{\cancel{5}^{1}} = \frac{45}{5}$$

$$x = 9$$

$$6x = 6 \cdot 9 = 54$$

7. Let x = one number and $x - 3$ = the other number. Then $x + x - 3 = 25$.

$$x + x - 3 = 25$$

$$2x - 3 = 25$$

$$2x - 3 + 3 = 25 + 3$$

$$2x = 28$$

$$\frac{\cancel{2}^1 x}{\cancel{2}} = \frac{28}{2}$$

$$x = 14$$

$$x - 3 = 11$$

8. Let x = the number of notebooks placed in one box and $x + 6$ = the number of notebooks placed in the other box. Then $x + x + 6 = 50$.

$$x + x + 6 = 50$$

$$2x + 6 = 50$$

$$2x + 6 - 6 = 50 - 6$$

$$2x = 44$$

$$\frac{\cancel{2}^1 x}{\cancel{2}^1} = \frac{44}{2}$$

$$x = 22$$

$$x + 6 = 22 + 6 = 28$$

9. Let x = the first integer, $x + 1$ = the second integer, and $x + 2$ = the third integer. Then $x + x + 1 + x + 2 = 81$.

$$3x + 3 = 81$$

$$3x + 3 - 3 = 81 - 3$$

$$3x = 78$$

$$\frac{\cancel{3}^1 x}{\cancel{3}^1} = \frac{78}{3}$$

$$x = 26$$

$$x + 1 = 26 + 1 = 27$$

$$x + 2 = 26 + 2 = 28$$

10. Let x = the first integer and $x + 2$ = the second integer. Then $x + x + 2 = 62$.

$$x + x + 2 = 62$$

$$2x + 2 = 62$$

$$2x + 2 - 2 = 62 - 2$$

$$2x = 60$$

$$\frac{\cancel{2}^1 x}{\cancel{2}^1} = \frac{60}{2}$$

$$x = 30$$

$$x + 2 = 32$$

In this section, you learned how to solve number problems. Each problem gives you the relationship between two or more numbers. From this information, you can set up an equation and solve for the numbers. Consecutive numbers increase by 1. Consecutive even numbers and consecutive odd numbers increase by 2.

Digit Problems

The symbols 0, 1, 2, 3, 4, 5, 6, 7, 8, and 9 are called *digits*. They are used to make our numbers. A number such as 28 is called a two-digit number. The eight is the ones digit and the two is the tens digit. The ones digit is also called the units digit. The number 28 means the sum of 2 tens and 8 ones and can be written as $2 \cdot 10 + 8 = 20 + 8$ or 28. The number 537 is called a three-digit number. The seven is the ones digit, the three is the tens digit, and the five is the hundreds digit. It can be written as $5 \cdot 100 + 3 \cdot 10 + 7$ or $500 + 30 + 7 = 537$.

A digit problem will sometimes ask you to find the sum of the digits. In order to do this, just add the digits. For example, the sum of the digits of the number 537 is $5 + 3 + 7 = 15$.

Sometimes digit problems will ask you to reverse the digits. If the digits of the number 36 or $(3 \cdot 10 + 6)$ are reversed, the new number is 63 or $(6 \cdot 10 + 3)$. Using this information and the material in the previous section, you will be able to solve digit problems.

EXAMPLE

The sum of the digits of a two-digit number is 9. If the digits are reversed, the new number is 63 more than the original number. Find the original number.

SOLUTION

Goal: You are being asked to find a certain two-digit number.

Strategy: Let $x =$ the tens digit and $9 - x =$ the ones digit. The original number can be written as $10x + (9 - x)$, and the number with the digits reversed can be written as $10(9 - x) + x$. Since the new number is 63 more than the original number, an equation can be written as

new number = original number + 63

$$10(9 - x) + x = 10x + (9 - x) + 63$$

Implementation: Solve the equation:

$$10(9-x) + x = 10x + (9-x) + 63$$
$$90 - 10x + x = 10x + 9 - x + 63$$
$$90 - 9x = 9x + 72$$
$$90 - 9x + 9x = 9x + 9x + 72$$
$$90 = 18x + 72$$
$$90 - 72 = 18x + 72 - 72$$
$$18 = 18x$$
$$\frac{18}{18} = \frac{\cancel{18}^{1}x}{\cancel{18}^{1}}$$
$$1 = x$$
$$9 - x = 9 - 1 = 8$$

The tens digit is 1 and the ones digit is 9 – 1 = 8. The number, then, is 18.

Evaluation: Take 18 and reverse the digits to get 81. Subtract 81 – 18 = 63. Hence the sum of the digits 1 + 8 is 9 and the difference of the two numbers is 63.

EXAMPLE

In a two-digit number, the tens digit is 3 less than the ones digit. If the digits of the number are reversed, the sum of the original number and the new number is 77. Find the original number.

SOLUTION

Goal: You are being asked to find a two-digit number.

Strategy: Let x = the ones digit and $x - 3$ = the tens digit. The number, then, is $10(x - 3) + x$. When the digits are reversed, the new number is $10x + (x - 3)$. Since their sum is 77, the equation is $10(x - 3) + x + 10x + x - 3 = 77$.

Implementation: Solve the equation:

$$10(x - 3) + x + 10x + x - 3 = 77$$
$$10x - 30 + x + 10x + x - 3 = 77$$
$$22x - 33 = 77$$

$$22x - 33 + 33 = 77 + 33$$
$$22x = 110$$
$$\frac{\cancel{22}^{1}x}{\cancel{22}_{1}} = \frac{110}{22}$$
$$x = 5$$

The ones digit is 5 and the tens digit is $x - 3 = 5 - 3 = 2$.

The number is 25.

Evaluation: The tens digit is 3 less than the ones digit. When the digits of 25 are reversed, the answer is 52; hence, $25 + 52 = 77$.

EXAMPLE

In a three-digit number, the ones digit is equal to the tens digit and the hundreds digit is 5 more than the ones digit. If the order of the digits is reversed, twice the new number is 268 less than the original number. Find the original number.

SOLUTION

Goal: You are being asked to find a three-digit number.

Strategy: Let $x =$ the ones digit and $x =$ the tens digit, since they are equal. The hundreds digit is $x + 5$ since it is 5 more than the ones digit. The number, then, is $100(x + 5) + 10x + x$. When the digits are reversed, the new number is 268 less than the original number. The equation is $100(x + 5) + 10x + x = 2(100x + 10x + x + 5) + 268$.

Implementation: Solve the equation:

$$100(x + 5) + 10x + x = 2(100x + 10x + x + 5) + 268$$
$$100x + 500 + 10x + x = 200x + 20x + 2x + 10 + 268$$
$$111x + 500 = 222x + 278$$
$$111x - 111x + 500 = 222x - 111x + 278$$
$$500 = 111x + 278$$
$$500 - 278 = 111x + 278 - 278$$
$$222x = 111x$$
$$\frac{222}{111} = \frac{\cancel{111}^{1}x}{\cancel{111}_{1}}$$
$$2 = x$$

Hence, the ones digit is 2, the tens digit is 2, and the hundreds digit is 2 + 5 = 7. The number is 722.

Evaluation: The number is 722, and reversing the digits, you get 227. Now 722 − 2 · 227 = 722 − 454 = 268.

TRY THESE

1. The sum of the digits of a two-digit number is 10. If 18 is added to the original number, the new number will have the same digits, but they will be reversed. Find the original number.

2. The sum of the digits of a two-digit number on a race car is 15. If the digits are reversed, the new number is 9 more than the original number. Find the original number.

3. If the digits of a two-digit number are reversed, the new number is 10 more than twice the original number. The sum of the digits of the original number is 8. Find the original number.

4. The tens digit of a two-digit number is 3 more than the ones digit. If the number is one more than eight times the sum of the digits, find the number.

5. In a two-digit number, the ones digit is 5 more than the tens digit. If the number is three times the sum of its digits, find the number.

6. In a two-digit number, the sum of the digits is 7. If the digits are reversed, three times the new number is 13 less than the original number. Find the original number.

7. The sum of the digits of a two-digit number on a football jersey is 11. If the tens digit is 3 more than the ones digit, find the number.

8. In a two-digit number, the ones digit is 4 less than the tens digit. The number is 3 less than seven times the sum of the digits. Find the number.

9. The sum of the digits in a two-digit number on a baseball jersey is 5. If the ones digit is 1 more than the tens digit, find the number.

10. In a two-digit number, the sum of the digits is 9. If the digits of the original number are reversed, the new number is 45 more than the original number. Find the original number.

SOLUTIONS

1. Let x be the tens digit and $10 - x$ be the ones digit. The number is $10x + 10 - x$. Reversing the digits, the new number is $10(10 - x) + x$. The equation is $10x + 10 - x + 18 = 10(10 - x) + x$.

$$10x + 10 - x + 18 = 10(10 - x) + x$$
$$9x + 28 = 100 - 10x + x$$
$$9x + 28 = 100 - 9x$$
$$9x + 9x + 28 = 100 - 9x + 9x$$
$$18x + 28 = 100$$
$$18x + 28 - 28 = 100 - 28$$
$$18x = 72$$
$$\frac{\cancel{18}^{1} x}{\cancel{18}^{1}} = \frac{72}{18}$$
$$x = 4$$
$$10 - x = 10 - 4 = 6$$

The number is 46.

2. Let x = the ones digit and $15 - x$ = the tens digit. The number is $10(15 - x) + x$. If the digits are reversed, the new number is $10x + 15 - x$. The equation is $10(15 - x) + x + 9 = 10x + 15 - x$.

$$10(15 - x) + x + 9 = 10x + 15 - x$$
$$150 - 10x + x + 9 = 10x + 15 - x$$
$$159 - 9x = 9x + 15$$
$$159 - 9x + 9x = 9x + 9x + 15$$
$$159 = 18x + 15$$
$$159 - 15 = 18x + 15 - 15$$
$$144 = 18x$$
$$\frac{144}{18} = \frac{\cancel{18}^{1} x}{\cancel{18}^{1}}$$
$$8 = x$$
$$15 - x = 15 - 8 = 7$$

Hence, the number is 78.

3. Let $x =$ the ones digit and $8 - x =$ the tens digit. The number is $10(8 - x) + x$. When the digits are reversed, the new number is $10x + 8 - x$. The equation is $2[10(8 - x) + x] + 10 = 10x + 8 - x$.

$$2[10(8 - x) + x] + 10 = 10x + 8 - x$$
$$20(8 - x) + 2x + 10 = 9x + 8$$
$$160 - 20x + 2x + 10 = 9x + 8$$
$$170 - 18x = 9x + 8$$
$$170 - 18x + 18x = 9x + 18x + 8$$
$$170 = 27x + 8$$
$$170 - 8 = 27x + 8 - 8$$
$$162 = 27x$$
$$\frac{162}{27} = \frac{\cancel{27}^{1}x}{\cancel{27}}$$
$$6 = x$$
$$8 - x = 8 - 6 = 2$$

The number is 26.

4. Let $x =$ the ones digits and $x + 3 =$ the tens digit. The number is $10(x + 3) + x$. The sum of the digits is $x + 3 + x$. The equation is $10(x + 3) + x = 8(x + 3 + x) + 1$.

$$10(x + 3) + x = 8(x + 3 + x) + 1$$
$$10x + 30 + x = 8x + 24 + 8x + 1$$
$$11x + 30 = 16x + 25$$
$$11x - 11x + 30 = 16x - 11x + 25$$
$$30 = 5x + 25$$
$$30 - 25 = 5x + 25 - 25$$
$$5 = 5x$$
$$\frac{\cancel{5}^{1}}{\cancel{5}^{1}} = \frac{\cancel{5}^{1}x}{\cancel{5}^{1}}$$
$$1 = x$$
$$x + 3 = 1 + 3 = 4$$

The number is 41.

5. Let x = the tens digit and $x + 5$ = the ones digit. The number is $10x + x + 5$. The sum of the digits is $x + x = 5$. The equation is $10x + x + 5 = 3(x + x + 5)$.

$$10x + x + 5 = 3(x + x + 5)$$
$$10x + x + 5 = 3x + 3x + 15$$
$$11x + 5 = 6x + 15$$
$$11x - 6x + 5 = 6x - 6x + 15$$
$$5x + 5 = 15$$
$$5x + 5 - 5 = 15 - 5$$
$$5x = 10$$
$$\frac{\cancel{5}^{1}x}{\cancel{5}^{1}} = \frac{10}{5}$$
$$x = 2$$
$$x + 5 = 2 + 5 = 7$$

The number is 27.

6. Let x = the ones digit and $7 - x$ = the tens digit. The number is $10(7 - x) + x$. When the digits are reversed, the new number is $10x + 7 - x$. The equation is $10(7 - x) + x = 3(10x + 7 - x) + 13$.

$$10(7 - x) + x = 3(10x + 7 - x) + 13$$
$$70 - 10x + x = 30x + 21 - 3x + 13$$
$$70 - 9x = 27x + 34$$
$$70 - 9x + 9x = 27x + 9x + 34$$
$$70 = 36x + 34$$
$$70 - 34 = 36x + 34 - 34$$
$$36 = 36x$$
$$\frac{36}{36} = \frac{\cancel{36}^{1}x}{\cancel{36}^{1}}$$
$$1 = x$$
$$7 - x = 7 - 1 = 6$$

The number is 61.

7. Let x = the ones digit and $x + 3$ = the tens digit. Since the sum of the digits is 11, the equation is $x + 3 + x = 11$.

$$x + 3 + x = 11$$
$$2x + 3 = 11$$

$$2x + 3 - 3 = 11 - 3$$
$$2x = 8$$
$$\frac{\cancel{2}^{1}x}{\cancel{2}^{1}} = \frac{8}{2}$$
$$x = 4$$
$$x + 3 = 4 + 3 = 7$$

The number is 74.

8. Let x = the ones digit and $x + 4$ = the tens digit. The number is $10(x + 4) + x$. The sum of the digits is $x + 4 + x$. The equation is $10(x + 4) + x + 3 = 7(x + 4 + x)$.

$$10(x + 4) + x + 3 = 7(x + 4 + x)$$
$$10x + 40 + x + 3 = 7x + 28 + 7x$$
$$11x + 43 = 14x + 28$$
$$11x - 11x + 43 = 14x - 11x + 28$$
$$43 = 3x + 28$$
$$43 - 28 = 3x + 28 - 28$$
$$15 = 3x$$
$$\frac{15}{3} = \frac{\cancel{3}^{1}x}{\cancel{3}^{1}}$$
$$5 = x$$
$$x + 4 = 5 + 4 = 9$$

The number is 95.

9. Let x = the ones digit and $5 - x$ = the tens digit. The equation is $x - 1 = 5 - x$.

$$x - 1 = 5 - x$$
$$x + x - 1 = 5 - x + x$$
$$2x - 1 = 5$$
$$2x - 1 + 1 = 5 + 1$$
$$2x = 6$$
$$\frac{\cancel{2}^{1}x}{\cancel{2}^{1}} = \frac{6}{2}$$
$$x = 3$$
$$5 - x = 5 - 3 = 2$$

The number is 23.

10. Let x = the ones digit and $9 - x$ = the tens digits. The number is $10(9 - x) + x$. When the digits are reversed, the new number is $10x + 9 - x$. The equation is $10(9 - x) + x + 45 = 10x + 9 - x$.

$$10(9 - x) + x + 45 = 10x + 9 - x$$
$$90 - 10x + x + 45 = 9x + 9$$
$$135 - 9x = 9x + 9$$
$$135 - 9x + 9x = 9x + 9x + 9$$
$$135 = 18x + 9$$
$$135 - 9 = 18x + 9 - 9$$
$$126 = 18x$$
$$\frac{126}{18} = \frac{\cancel{18}^{1}x}{\cancel{18}^{1}}$$
$$7 = x$$
$$9 - x = 9 - 7 = 2$$

The number is 27.

This section explained how to solve problems involving digits. Our number system consists of 10 digits, and each number consists of a ones digit, a tens digit, a hundreds digit, etc.

Summary

This chapter explained two types of problems, namely number problems and digit problems. Although both sections use numbers, the equations to solve the problems are somewhat different. It is necessary to be aware of the difference.

In number problems, sometimes you are looking for one number, two numbers, or three numbers. Sometimes the numbers are consecutive numbers, or consecutive odd and even numbers.

In digit problems, you are usually looking for the relationship between the digits of a single number.

QUIZ

1. **If the sum of three consecutive numbers is 39, the largest of the three numbers is**
   A. 12
   B. 14
   C. 118
   D. 13

2. **If one number is six times another number and the sum of the numbers is 147, the smaller number is**
   A. 21
   B. 14
   C. 28
   D. 31

   882 ✗✗✗ × ≠ 147 ✗

3. **The sum of two numbers is 39 and one number is 5 more than the other number. The smaller number is**
   A. 22
   B. 12
   C. 17
   D. 14

4. **If the sum of two consecutive even numbers is 166, the smaller number is**
   A. 84
   B. 86
   C. 80
   D. 82

5. **If the sum of two consecutive numbers is 157, the smaller number is**
   A. 84
   B. 63
   C. 78
   D. 61

6. **The tens digit of a two digit number is 4 more than the ones digit of a two-digit number. If 13 is added to the number, the answer is 75. Find the number.**
   A. 51
   B. 62
   C. 15
   D. 46

7. The sum of the digits of a two-digit number is 12. If the digits are reversed, the new number is 36 more than the original number. Find the number.

   A. 48
   B. 57
   C. 66
   D. 39

8. In a two-digit number, the ones digit is 4 more than the tens digit. Three times the number is 74 more than the number. Find the number.

   A. 26
   B. 59
   C. 15
   D. 37

9. In a two-digit number, the tens digit is 5 less than the ones digit. If the sum of the digits is 27 less than the original number, find the original number.

   A. 27
   B. 38
   C. 49
   D. 61

10. The sum of the digits in a two-digit number is 9. If the digits are reversed, the new number is 18 more than twice the original number. Find the original number.

    A. 27
    B. 45
    C. 36
    D. 72

chapter 7

Solving Coin and Age Problems

This chapter explains how to solve coin and age problems. Coin problems consist of problems about metal coins such as pennies, nickels, dimes, etc., but could also include paper money or stamps. Any problems in which a money value can be assigned to objects can be solved using the same techniques as solving a coin problem. Coin problems can be solved by using the values of the coins, then setting up and solving the equation.

Age problems usually include finding the ages of two people, such as the age of a father and his daughter. Sometimes age problems include the person's present age and his past or future age. The equation is set up using this information.

CHAPTER OBJECTIVES

In this chapter, you will learn how to

- Solve problems involving coins
- Solve age problems

Coin Problems

Suppose you have some coins in your pocket or wallet. In order to determine the amount of money you have, you multiply the value of each type of coin by the number of coins of that denomination and then add the answers. For example, if you have six nickels, four dimes, and two quarters, the total amount of money you have in change is

$$6 \times 5¢ + 4 \times 10¢ + 2 \times 25¢ =$$
$$30¢ \quad + \quad 40¢ \quad + \quad 50¢ \quad = 120¢ \text{ or } \$1.20.$$

In general, then, to find the amount of money for

Pennies—multiply the number of pennies by 1¢

Nickels—multiply the number of nickels by 5¢

Dimes—multiply the number of dimes by 10¢

Quarters—multiply the number of quarters by 25¢

Half dollars—multiply the number of half dollars by 50¢

To solve problems involving coins:

1. Let x = the number of one type of coin (i.e., pennies, nickels, dimes, etc.). Write the numbers of the other type of coins in terms of x.
2. Set up the equation by multiplying the number of each type of coin by the value of the coin.
3. Solve the equation for x, then find the other numbers.
4. Check the answers.

If you want to avoid decimals, you can work with cents rather than dollars. You can change dollars to cents by multiplying by 100. You can change the answer back to dollars by dividing by 100.

EXAMPLE

A person has 16 coins consisting of quarters and nickels. If the total amount of this change is $2.60, how many of each kind of coin are there?

SOLUTION

Goal: You are being asked to find the number of quarters and the number of nickels the person has.

Strategy: Let x = the number of quarters and $(16 - x)$ = the number of nickels; then the value of the quarters is $25x$ and the value of the nickels is $5(16 - x)$. The total amount of money in cents is $\$2.60 \cdot 100 = 260$. The equation is $25x + 5(16 - x) = 260$.

Implementation: Solve the equation:

$$25x + 5(16 - x) = 260$$
$$25x + 80 - 5x = 260$$
$$20x + 80 = 260$$
$$20x + 80 - 80 = 260 - 80$$
$$20x = 180$$
$$\frac{\cancel{20}^{1} x}{\cancel{20}^{1}} = \frac{180}{20}$$
$$x = 9 \qquad \text{(quarters)}$$
$$16 - x = 16 - 9 = 7 \qquad \text{(nickels)}$$

There are 9 quarters and 7 nickels.

Evaluation: The value of 9 quarters and 7 nickels is $9 \times 25¢ + 7 \times 5¢ = 225 + 35 = 260¢ = \$2.60$.

EXAMPLE

A person has five times as many pennies as he has dimes and eight more nickels than dimes. If the total amount of these coins is \$1, how many of each kind of coin does he have?

SOLUTION

Goal: You are being asked to find the number of nickels, pennies, and dimes.

Strategy: Let x = the number of dimes, $5x$ = the number of pennies, and $x + 8$ = the number of nickels. Then the value of the dimes is $10x$, the value of the pennies is $1 \cdot 5x$, and the value of the nickels is $5(x + 8)$. The total amount is $\$1 \times 100$ or $100¢$. The equation is $10x + 1 \cdot 5x + 5(x + 8) = 100$.

Implementation: Solve the equation:

$$10x + 1 \cdot 5x + 5(x + 8) = 100$$

$$10x + 5x + 5x + 40 = 100$$

$$20x + 40 = 100$$

$$20x + 40 - 40 = 100 - 40$$

$$20x = 60$$

$$\frac{\cancel{20}^{1} x}{\cancel{20}_{1}} = \frac{60}{20}$$

$$x = 3 \qquad \text{(dimes)}$$

$$5x = 5 \cdot 3 = 15 \qquad \text{(pennies)}$$

$$x + 8 = 3 + 8 = 11 \qquad \text{(nickels)}$$

There are 3 dimes, 15 pennies, and 11 nickels.

Evaluation: The value of 3 dimes, 15 pennies, and 11 nickels is $3 \times 10¢ + 15 \times 1¢ + 11 \times 5¢ = 30 + 15 + 55 = \$1.00$.

Other types of problems involving values can be solved using the same strategy as the coin problems. Consider the next example.

EXAMPLE

A person bought 10 candy bars consisting of caramel twists costing $0.88 each and chocolate marshmallow bars costing $1.19 each. If the total cost of the candy was $10.97, find the number of each kind of candy bar the person bought.

SOLUTION

Goal: You are being asked to find how many caramel twists and how many chocolate marshmallow candy bars the person bought.

Strategy: Let $x =$ the number of caramel twists and $(10 - x) =$ the number of chocolate marshmallow bars. Since the caramel twists cost $0.88 each, the value of the caramel twists is $0.88x$, and since the chocolate marshmallow bars costs $1.19 each, the value of the chocolate marshmallow bars is $\$1.19(10 - x)$. The equation is $0.88x + 1.19(10 - x) = 10.97$.

Implementation: Solve the equation:

$$0.88x + 1.19(10 - x) = 10.97$$

$$0.88x + 11.9 - 1.19x = 10.97$$

$$11.9 - 0.31x = 10.97$$

$$11.9 - 11.9 - 0.31x = 10.97 - 11.9$$

$$-0.31x = -0.93$$

$$\frac{\cancel{-0.31}^{1}x}{\cancel{-0.31}^{1}x} = \frac{-0.93}{-0.31}$$

$$x = 3 \qquad \text{(caramel twists)}$$

$$10 - x = 10 - 3 = 7 \quad \text{(chocolate marshmallow bars)}$$

The person bought three caramel twists and seven chocolate marshmallow bars.

Evaluation: Three caramel twists and seven chocolate marshmallow bars cost $3 \times \$0.88 + 7 \times \$1.19 = 2.64 + 8.33 = \$10.97$.

Still Struggling

The preceding problem was worked out in dollars rather than in cents. Either way is correct. The equation in cents would be $88x + 119(10 - x) = 1,097$.

TRY THESE

1. A person has twice as many dimes as she has nickels and three more nickels than pennies. If the total amount of the coins is $1.01, find the number of each type of coin the person has.

2. A person has five more quarters than pennies. If the total amount of the coins is $2.29, find the number of pennies and quarters the person has.

3. A person bought 10 stamps consisting of 44¢ stamps and 50¢ stamps. If the cost of the stamps is $4.64, find the number of each type of the stamps purchased.

4. If a person has four times as many nickels as quarters and the total amount of money is $1.35, find the number of quarters and nickels.

5. A drug store sells a bottle of vitamin C for $3.75 and a bottle of vitamin E for $6.29. If a person purchased three bottles and paid $13.79, how many bottles of each vitamin did the person purchase?

6. A dairy store sold a total of 63 snow cones and popsicles. If the snow cones cost $1.25 each and the popsicles cost $0.75 each and the store made $62.25, find the number of each sold.

7. In a child's savings bank, there are five times as many quarters as half dollars and eight more dimes than half dollars. If the total amount of the money in the bank is $6.35, find the number of each type of coin in the bank.

8. A clerk is given $120 in bills to put in a cash drawer at the start of a workday. There are three times as many $1 bills as $5 bills and six fewer $10 bills than $5 bills. How many of each type of bill are there?

9. A child's bank contains 29 coins consisting of nickels and dimes. If the total amount of money is $1.90, find the number of nickels and dimes in the bank.

10. A pile of 24 coins consists of dimes and nickels. If the total amount of the coins is $1.50, find the number of dimes and nickels.

SOLUTIONS

1. Let $x =$ the number of nickels, $2x =$ the number of dimes, and $x - 3 =$ the number of pennies. The value of the dimes is $10 \cdot 2x = 20x$, the value of the nickels is $5x$, and the value of the pennies is $1 \cdot (x - 3)$. The equation is $20x + 5x + (x - 3) = 101$.

$$20x + 5x + (x - 3) = 101$$
$$20x + 5x + x - 3 = 101$$
$$26x - 3 = 101$$
$$26x - 3 + 3 = 101 + 3$$
$$26x = 104$$

$$\frac{\cancel{26}^{1} x}{\cancel{26}_{1}} = \frac{104}{26}$$

$$x = 4 \qquad \text{(nickels)}$$
$$2x = 2 \cdot 4 = 8 \qquad \text{(dimes)}$$
$$x - 3 = 4 - 3 = 1 \qquad \text{(penny)}$$

2. Let x = the number of pennies and $x + 5$ = the number of quarters. The value of the pennies is $1x$ and the value of the quarters is $25(x + 5)$. The equation is $x + 25(x + 5) = 229$.

$$x + 25(x + 5) = 229$$
$$x + 25x + 125 = 229$$
$$26x + 125 = 229$$
$$26x + 125 - 125 = 229 - 125$$
$$26x = 104$$
$$\frac{\overset{1}{\cancel{26}}\,x}{\underset{1}{\cancel{26}}} = \frac{104}{26}$$

$$x = 4 \qquad \text{(pennies)}$$
$$x + 5 = 4 + 5 = 9 \qquad \text{(quarters)}$$

There are four pennies and nine quarters.

3. Let x = the number of 44¢ stamps and $10 - x$ = the number of 50¢ stamps. The value of the 44¢ stamps is $44x$ and the value of the 50¢ stamps is $50(10 - x)$. The equation is $44x + 50(10 - x) = 464$.

$$44x + 50(10 - x) = 464$$
$$44x + 500 - 50x = 464$$
$$-6x + 500 = 464$$
$$-6x + 500 - 500 = 464 - 500$$
$$-6x = -36$$
$$\frac{\overset{1}{\cancel{-6}}\,x}{\underset{1}{\cancel{-6}}} = \frac{-36}{-6}$$

$$x = 6$$
$$10 - x = 10 - 6 = 4$$

There are six 44¢ stamps and four 50¢ stamps.

4. Let x = the number of quarters and $4x$ = the number of nickels. The value of the quarters is $25x$ and the value of the nickels is $5 \cdot 4x$ or $20x$. The equation is $25x + 20x = 135$.

$$25x + 20x = 135$$
$$45x = 135$$

$$\frac{\cancel{45}^{1}\,x}{\cancel{45}_{1}} = \frac{135}{45}$$

$$x = 3 \qquad \text{(quarters)}$$

$$4x = 4 \cdot 3 = 12 \qquad \text{(nickels)}$$

There are 3 quarters and 12 nickels.

5. Let x = the number of bottles of vitamin C and $3 - x$ = the number of bottles of vitamin E. The vitamin C costs $375x$, and the vitamin E costs $629(3 - x)$. The equation is $375x + 629(3 - x) = 1{,}379$

$$375x + 629(3 - x) = 1{,}379$$

$$375x + 1{,}887 - 629x = 1{,}379$$

$$-254x + 1{,}887 = 1{,}379$$

$$-254x + 1{,}887 - 1{,}887 = 1{,}379 - 1{,}887$$

$$-254x = -508$$

$$\frac{\cancel{-254}^{1}\,x}{\cancel{-254}} = \frac{-508}{-254}$$

$$x = 2 \qquad \text{(bottles of vitamin C)}$$

$$3 - x = 3 - 2 = 1 \qquad \text{(bottle of vitamin E)}$$

There are two bottles of vitamin C and one bottle of vitamin E.

6. Let x = the number of snow cones sold and $63 - x$ = the number of popsicles sold. The snow cones cost $125x$ and the popsicles cost $75(63 - x)$. The equation is $125x + 75(63 - x) = 6{,}225$.

$$125x + 75(63 - x) = 6{,}225$$

$$125x + 4{,}725 - 75x = 6{,}225$$

$$50x + 4{,}725 = 6{,}225$$

$$50x + 4{,}725 - 4{,}725 = 6{,}225 - 4{,}725$$

$$50x = 1{,}500$$

$$\frac{\cancel{50}^{1}\,x}{\cancel{50}_{1}} = \frac{1{,}500}{50}$$

$$x = 30 \qquad \text{(snow cones)}$$

$$63 - x = 63 - 30 = 33 \qquad \text{(popsicles)}$$

The store sold 30 snow cones and 33 popsicles.

7. Let x = the number of half dollars, $5x$ = the number of quarters, and $x + 8$ = the number of dimes. The value of the half dollars is $50x$. The value of the quarters is $25 \cdot 5x = 125x$, and the value of the dimes is $10(x + 8)$. The equation is $50x + 125x + 10(x + 8) = 635$.

$$50x + 125x + 10(x + 8) = 635$$

$$50x + 125x + 10x + 80 = 635$$

$$185x + 80 = 635$$

$$185x + 80 - 80 = 635 - 80$$

$$185x = 555$$

$$\frac{\cancel{185}^{1} x}{\cancel{185}_{1}} = \frac{555}{185}$$

$$x = 3 \qquad \text{(half dollars)}$$

$$5x = 5 \cdot 3 = 15 \qquad \text{(quarters)}$$

$$x + 8 = 3 + 8 = 11 \qquad \text{(dimes)}$$

There are 3 half dollars, 15 quarters, and 11 dimes.

8. Let x = the number of \$5 bills. Let $3x$ = the number of \$1 bills, and $x - 6$ = the number of \$10 bills. The value of the \$5 bills is $5x$. The value of the \$1 bills is $1 \cdot 3x$, and the value of the \$10 bills is $10(x - 6)$. The equation is $5x + 3x + 10(x - 6) = 120$.

$$5x + 3x + 10(x - 6) = 120$$

$$5x + 3x + 10x - 60 = 12$$

$$18x - 60 = 120$$

$$18x - 60 + 60 = 120 + 60$$

$$18x = 180$$

$$\frac{\cancel{18}^{1} x}{\cancel{18}_{1}} = \frac{180}{18}$$

$$x = 10 \qquad \text{(the number of \$5 bills)}$$

$$3x = 3 \cdot 10 = 30 \qquad \text{(the number of \$1 bills)}$$

$$x - 6 = 10 - 6 = 4 \qquad \text{(the number of \$10 bills)}$$

There are 10 \$5 bills, 30 \$1 bills, and 4 \$10 bills.

9. Let x = the number of nickels and $29 - x$ = the number of dimes. The value of the nickels is $5x$, and the value of the dimes is $10(29 - x)$. The equation is $5x + 10(29 - x) = 190$.

$$5x + 10(29 - x) = 190$$

$$5x + 290 - 10x = 190$$

$$-5x + 290 = 190$$

$$-5x + 290 - 290 = 190 - 290$$

$$-5x = -100$$

$$\frac{\cancel{-5}^{1} x}{\cancel{-5}_{1}} = \frac{-100}{-5}$$

$$x = 20 \qquad \text{(nickels)}$$

$$29 - x = 29 - 20 = 9 \qquad \text{(dimes)}$$

There are 20 nickels and 9 dimes.

10. Let x = the number of dimes and $24 - x$ = the number of nickels. The value of the dimes is $10x$ and the value of the nickels is $5(24 - x)$. The equation is $10x + 5(24 - x) = 150$.

$$10x + 5(24 - x) = 150$$

$$10x + 120 - 5x = 150$$

$$5x + 120 = 150$$

$$5x + 120 - 120 = 150 - 120$$

$$5x = 30$$

$$\frac{\cancel{5}^{1} x}{\cancel{5}_{1}} = \frac{30}{5}$$

$$x = 6 \qquad \text{(dimes)}$$

$$24 - x = 24 - 6 = 18 \qquad \text{(nickels)}$$

There are 6 dimes and 18 nickels.

In this section, you learned how to solve coin problems. The technique is to set up the equation by representing the number of coins using x and multiplying each number of coins by their numerical values: 1¢ for pennies, 5¢ for nickels, 10¢ for dimes, 25¢ for quarters, and 50¢ for half dollars.

Age Problems

When you encounter an age problem, you will often see that the problem gives you information about the age of a person in the future or in the past. For example, if a mother is three times as old as her daughter, their present ages can be represented as

Let x = the daughter's age and

$3x$ = the mother's age

Now, if the problem gives you information about their ages seven years from now, you can represent their future ages as

Let $x + 7$ = the daughter's future age and

$3x + 7$ = the mother's future age

Likewise, if the problem gives you some information about their ages, 10 years ago, you can represent their past ages as

Let $x - 10$ = the daughter's past age and

$3x - 10$ = the mother's past age

The basic strategy for solving age problems is to represent the present ages of the people, represent the past or future ages of the people, and then set up the equation and solve it.

EXAMPLE

A father is six times as old as his son; in 20 years, he will be twice as old as his son. Find their present ages.

SOLUTION

Goal: You are being asked to find the present ages of the father and his son.

Strategy: Let x = the son's present age and $6x$ = the father's present age. In 20 years, the son's age will be $x + 20$ and the father's age will be $6x + 20$. If the father will be twice as old as his son in 20 years, the equation is two times the son's age in 20 years = the father's age in 20 years or $2(x + 20) = 6x + 20$.

Implementation: Solve the equation:

$$2(x + 20) = 6x + 20$$

$$2x + 40 = 6x + 20$$

$$2x - 2x + 40 = 6x - 2x + 20$$

$$40 = 4x + 20$$

$$40 - 20 = 4x + 20 - 20$$

$$20 = 4x$$

$$\frac{20}{4} = \frac{\cancel{4}^1 x}{\cancel{4}^1}$$

$$5 = x \text{ or } x = 5 \qquad \text{(son's age)}$$

$$6 \cdot x = 6 \cdot 5 = 30 \qquad \text{(father's age)}$$

Evaluation: In 20 years, the son's age will be $5 + 20 = 25$ and the father's age will be $30 + 20 = 50$. Since $50 = 2 \cdot 25$, the father will be twice as old as the son.

EXAMPLE

Eli is nine years older than his sister. In six years, Eli will be twice as old as his sister. Find their present ages.

SOLUTION

Goal: You are being asked to find the present ages of Eli and his sister.

Strategy: Let $x =$ Eli's sister's age and $x + 9 =$ Eli's age. In six years, their ages will be $x + 6 =$ Eli's sister's age and $(x + 9) + 6 =$ Eli's age. In six years, Eli will be twice as old means two times Eli's sister's age in six years = Eli's age in 6 years or $2(x + 6) = (x + 9) + 6$.

Implementation: Solve the equation:

$$2(x + 6) = (x + 9) + 6$$

$$2x + 12 = x + 9 + 6$$

$$2x + 12 = x + 15$$

$$2x - x + 12 = x - x + 15$$

$$x + 12 = 15$$

$$x + 12 - 12 = 15 - 12$$

$$x = 3 \qquad \text{(Eli's sister's age)}$$

$$x + 9 = 3 + 9 = 12 \qquad \text{(Eli's age)}$$

Evaluation: In six years, Eli's sister will be $3 + 6 = 9$ years, and Eli will be $12 + 6 = 18$, which is twice his sister's age.

EXAMPLE

Sarah is 11 years older than Beth. If the sum of their ages is 67, find each one's age.

SOLUTION

Goal: You are being asked to find the ages of Sarah and Beth.

Strategy: Let $x =$ Beth's age and $x + 11 =$ Sarah's age. Then the sum of their ages is $x + x + 11 = 67$.

Implementation: Solve the equation:

$$x + x + 11 = 67$$
$$2x + 11 = 67$$
$$2x + 11 - 11 = 67 - 11$$
$$2x = 56$$
$$\frac{\cancel{2}^{1} x}{\cancel{2}^{1}} = \frac{56}{2}$$
$$x = 28 \qquad \text{(Beth's age)}$$
$$x + 11 = 28 + 11 = 39 \qquad \text{(Sarah's age)}$$

Evaluation: Sarah is 11 years older than Beth, and the sum of their ages is $39 + 28 = 67$.

EXAMPLE

A mother is 36 years old and her daughter is 14 years old. In how many years will the mother be twice as old as her daughter?

SOLUTION

Goal: You are being asked to find the number of years it will be until the mother is twice as old as her daughter.

Strategy: Let $x =$ the number of years. Then the mother's age in x years will be $36 + x$ years, and the daughter's age in x years will be $14 + x$ years. If the mother is twice as old as the daughter in x years, the equation is $2(14 + x) = 36 + x$.

Implementation: Solve the equation:

$$2(14 + x) = 36 + x$$
$$28 + 2x = 36 + x$$
$$28 + 2x - x = 36 + x - x$$
$$28 + x = 36$$
$$28 - 28 + x = 36 - 28$$
$$x = 8$$

Hence, in eight years the mother will be twice as old as her daughter.

Evaluation: In eight years, the mother will be $36 + 8 = 44$ years old, and the daughter will be $14 + 8 = 22$ years, in which case the mother will be twice as old as her daughter.

TRY THESE

1. Mike is 32 and Joan is 22. How many years ago was Mike twice as old as Joan?

2. Beth is eight years older than Megan. Eleven years ago, Beth was three times as old as Megan. Find their present ages.

3. The sum of Judy and Sam's ages is 66. Judy was twice as old as Sam 15 years ago. Find their present ages.

4. A father is three times as old as his daughter. Fifteen years ago, he was nine times as old as his daughter. How old are they now?

5. A woman is five times as old as her neighbor's son. In 24 years, she will be twice as old as the son. How old are they now?

6. Bart is three years older than Bret. In seven years, Bart will be twice as old as Bret was one year ago. Find their present ages.

7. A father is four times as old as his twin sons. If the sum of their ages in three years will be 75, how old are they now?

8. Cindy is six years older than Mindy. Four years from now, Cindy will be three times as old as Mindy was two years ago. Find their present ages.

9. Sid is five years older than his brother Tim. If the sum of their ages is 37, how old are they now?

10. Pam is eight years older than her brother. Four years from now, the sum of their ages will be 30. Find their present ages.

SOLUTIONS

1. Let x = the number of years ago that Mike was twice as old as Joan. $32 - x$ was Mike's age at that time, and $22 - x$ was Joan's age at that time. Since Mike was twice as old as Joan, the equation is $32 - x = 2(22 - x)$.

$$32 - x = 2(22 - x)$$
$$32 - x = 44 - 2x$$
$$32 - x + 2x = 44 - 2x + 2x$$
$$32 + x = 44$$
$$32 - 32 + x = 44 - 32$$
$$x = 12$$

Hence, 12 years ago, Mike was twice as old as Joan. That is, Mike was $32 - 12 = 20$ years old and Joan was $22 - 12 = 10$ years old.

2. Let x = Megan's age and $x + 8$ = Beth's age. Eleven years ago, Megan's age was $x - 11$ and Beth's age was $x + 8 - 11 = x - 3$. At that time, Beth was three times as old as Megan, so the equation is $x - 3 = 3(x - 11)$.

$$x - 3 = 3(x - 11)$$
$$x - 3 = 3x - 33$$
$$x - 3x - 3 = 3x - 3x - 33$$
$$-2x - 3 = -33$$
$$-2x - 3 + 3 = -33 + 3$$
$$-2x = -30$$
$$\frac{\cancel{-2}^{1}x}{\cancel{-2}_{1}} = \frac{-30}{-2}$$
$$x = 15 \qquad \text{(Megan's age)}$$
$$x + 8 = 15 + 8 = 23 \qquad \text{(Beth's age)}$$

3. Let x = Judy's age and $66 - x$ = Sam's age. Fifteen years ago, Judy's age would have been $x - 15$ and Sam's age would have been $66 - x - 15$ or $51 - x$. Since Judy was twice as old as Mike, the equation is $x - 15 = 2(51 - x)$.

$$x - 15 = 2(51 - x)$$
$$x - 15 = 102 - 2x$$
$$x + 2x - 15 = 102 - 2x + 2x$$

$$3x - 15 = 102$$

$$3x - 15 + 15 = 102 + 15$$

$$3x = 117$$

$$\frac{\cancel{3}^{1} x}{\cancel{3}^{1}} = \frac{117}{3}$$

$$x = 39 \qquad \text{(Judy's age)}$$

$$66 - x = 66 - 39 = 27 \qquad \text{(Sam's age)}$$

4. Let x = the daughter's age and $3x$ = the father's age. Fifteen years ago, the daughter's age was $x - 15$, and the father's age was $3x - 15$. Since the father was nine times as old as the daughter, the equation is $3x - 15 = 9(x - 15)$.

$$3x - 15 = 9(x - 15)$$

$$3x - 15 = 9x - 135$$

$$3x - 9x - 15 = 9x - 9x - 135$$

$$-6x - 15 = -135$$

$$-6x - 15 + 15 = -135 + 15$$

$$\frac{\cancel{-6}^{1} x}{\cancel{-6}^{1}} = \frac{-120}{-6}$$

$$x = 20 \qquad \text{(daughter's age)}$$

$$3x = 3(20) = 60 \qquad \text{(father's age)}$$

5. Let x = the son's age and $5x$ = the woman's age. In 24 years, the son will be $x + 24$ years old, and the woman will be $5x + 24$ years old. Since she will be twice as old as the son, the equation is $5x + 24 = 2(x + 24)$.

$$5x + 24 = 2(x + 24)$$

$$5x + 24 = 2x + 48$$

$$5x - 2x + 24 = 2x - 2x + 48$$

$$3x + 24 = 48$$

$$3x + 24 - 24 = 48 - 24$$

$$3x = 24$$

$$\frac{\cancel{3}^1 x}{\cancel{3}^1} = \frac{24}{3}$$

$$x = 8 \qquad \text{(son's age)}$$

$$5x = 5(8) = 40 \qquad \text{(woman's age)}$$

6. Let x = Bret's age and x + 3 = Bart's age. In seven years, Bart's age will be $x + 3 + 7$ or $x + 10$. Since Bart will be twice as old as Bret was one year ago, the equation is $x + 10 = 2(x - 1)$.

$$x + 10 = 2(x - 1)$$

$$x + 10 = 2x - 2$$

$$x - 2x + 10 = 2x - 2x - 2$$

$$-x + 10 = -2$$

$$-x + 10 - 10 = -2 - 10$$

$$-x = -12$$

$$\frac{\cancel{-1}^1 x}{\cancel{-1}^1} = \frac{-12}{-1}$$

$$x = 12 \qquad \text{(Bret's age)}$$

$$x + 3 = 12 + 3 = 15 \qquad \text{(Bart's age)}$$

7. Let x = each twin's age and 4x = the father's age; in three years, each twin will be x + 3 years old and the father's age will be 4x + 3. Since the sum of their ages is 75, the equation is $x + 3 + x + 3 + 4x + 3 = 75$.

$$x + 3 + x + 3 + 4x + 3 = 75$$

$$6x + 9 = 75$$

$$6x + 9 - 9 = 75 - 9$$

$$6x = 66$$

$$\frac{\cancel{6}^1 x}{\cancel{6}^1} = \frac{66}{6}$$

$$x = 11 \qquad \text{(each twin's age)}$$

$$4x = 4(11) = 44 \qquad \text{(father's age)}$$

8. Let x = Mindy's age and $x + 6$ = Cindy's age. In four years, Cindy will be $x + 6 + 4 = x + 10$ years old, and Mindy's age two years ago was $x - 2$. Since Cindy will be three times as old as Mindy was two years ago, the equation is $x + 10 = 3(x - 2)$.

$$x + 10 = 3(x - 2)$$

$$x + 10 = 3x - 6$$

$$x - 3x + 10 = 3x - 3x - 6$$

$$-2x + 10 = -6$$

$$-2x + 10 - 10 = -6 - 10$$

$$-2x = -16$$

$$\frac{\cancel{-2}^{1} x}{\cancel{-2}^{1}} = \frac{-16}{-2}$$

$$x = 8 \qquad \text{(Mindy's age)}$$

$$x + 6 = 8 + 6 = 14 \qquad \text{(Cindy's age)}$$

9. Let x = Tim's age and $x + 5$ = Sid's age. If the sum of their ages is 37, then the equation is $x + x + 5 = 37$.

$$x + x + 5 = 37$$

$$2x + 5 = 37$$

$$2x + 5 - 5 = 37 - 5$$

$$2x = 32$$

$$\frac{\cancel{2}^{1} x}{\cancel{2}^{1}} = \frac{32}{2}$$

$$x = 16 \qquad \text{(Tim's age)}$$

$$x + 5 = 16 + 5 = 21 \qquad \text{(Sid's age)}$$

10. Let Pam's age = $x + 8$ and her brother's age = x. In four years, Pam will be $x + 8 + 4 = x + 12$ years old, and her brother will be $x + 4$ years old. The equation is $x + 4 + x + 12 = 30$, since the sum of their ages will be 30.

$$x + 4 + x + 12 = 30$$

$$2x + 16 = 30$$

$$2x + 16 - 16 = 30 - 16$$

$$2x = 14$$

$$\frac{\cancel{2}^{1} x}{\cancel{2}^{1}} = \frac{14}{2}$$

$$x = 7 \qquad \text{(brother's age)}$$

$$x + 8 = 7 + 8 = 15 \qquad \text{(Pam's age)}$$

Summary

In this section, you have learned how to solve age problems. The key to the solution is to let $x =$ one person's age, then represent the other person's age in terms of x. Set up the equation using both ages and the condition or conditions given in the problem and solve. Be sure to check your answers.

This chapter explained how to solve coin and age problems.

QUIZ

1. A person has three more quarters than nickels. If the total amount of money is $1.95, find the number of nickels the person has.

   A. 2

   B. 4

   C. 6

   D. 8

2. A person has 11 coins in his pocket consisting of dimes and quarters. If he has one more quarter than dimes and a total of $2, how many dimes does he have?

   A. 6

   B. 4

   C. 3

   D. 5

3. A money box contains six more pennies than nickels and seven more dimes than nickels. If the total amount of money in the bank is $1.72, find the number of dimes in the bank.

   A. 4

   B. 6

   C. 13

   D. 12

4. A person has 15 bills consisting of $1 bills and $5 bills. If the total amount of money the person has is $43, find the number of $5 bills the person has.

   A. 5

   B. 7

   C. 10

   D. 15

5. A person has twice as many pennies as he has quarters, and he has five fewer dimes as he has pennies. If he has a total of $1.38, how many pennies does he have?

   A. 8

   B. 3

   C. 4

   D. 6

6. The sum of Bill and Lonny's ages is 52. Six years ago, Bill was three times as old as Lonny. Find Bill's present age.

   A. 32

   B. 36

   C. 4

   D. 6

7. **Bob's brother is 10 years older than Bob. If the sum of their ages is 26, find Bob's age.**
   A. 10
   B. 15
   C. 8
   D. 6

8. **Mary is four times as old as her younger sister. In 10 years, she will be twice as old as her sister. How old is Mary today?**
   A. 15
   B. 20
   C. 25
   D. 30

9. **The sum of Brooke's age and her best friend's age is 51, and the difference in their ages is 3. Brooke is the older. How old is Brooke?**
   A. 24
   B. 25
   C. 26
   D. 27

10. **Carrie is three years younger than her husband. In seven years, the sum of their ages will be 101. How old is Carrie?**
    A. 36
    B. 50
    C. 32
    D. 42

chapter **8**

Solving Distance and Mixture Problems

This chapter explains how to solve distance problems and mixture problems. Distance problems usually involve two vehicles moving either in the same direction or opposite directions and at different speeds. Also, these problems could include boats traveling up and down a stream taking into account the speed of the current, or airplanes flying with or against the wind.

Mixture problems involve mixing two solutions or solids to get a third mixture consisting of both items. Mixture problems could also include diluting solutions—that is, making them weaker.

CHAPTER OBJECTIVES

In this chapter, you will learn how to

- Solve distance problems
- Solve mixture problems

Distance Problems

The basic formula for solving distance problems is Distance = Rate × Time or $D = RT$. For example, if an automobile travels at 30 miles per hour for 2 hours, then the distance is $D = RT = 30 \times 2 = 60$ miles.

Distance problems usually involve two vehicles (i.e., automobiles, trains, bicycles, etc.) either traveling in the same direction or in opposite directions or one vehicle making a round trip. The procedure for solving distance problems is

1. Draw a diagram of the situation.

2. Set up a table as shown.

	Rate	×	Time	=	Distance
First vehicle					
Second vehicle					

3. Fill in the information in the table.

4. Write an equation for the situation, and solve it.

EXAMPLE

A person rode his bike on a bike trail at a rate of 10 miles per hour. While on the trail, he had a flat tire and had to walk back to his automobile at a rate of 2 miles per hour. If the total time he traveled was 2.4 hours, how far did he ride?

SOLUTION

Goal: You are being asked to find the distance that the person rode until he got a flat tire.

Strategy: The distance he walked and rode are the same, but the directions are different. See Figure 8-1.

| Automobile | Distance rode | Flat time |
| Automobile | Distance walked | Flat time |

FIGURE 8-1

Place the rate for riding 10 miles per hour and rate for walking 2 miles per hour in the boxes under Rate. Let t be the time he rode and $2.4 - t$ be the time that he walked. Place these in the boxes under Time. To get the distance, multiply the rates by the times and place these expressions in the boxes under Distance, as shown.

	Rate	×	Time	=	Distance
Riding	10		t		$10t$
Walking	2		$2.4 - t$		$2(2.4 - t)$

Since the distances are equal, the equation is $10t = 2(2.4 - t)$.

Implementation: Solve the equation:

$$10t = 2(2.4 - t)$$
$$10t = 4.8 - 2t$$
$$10t + 2t = 4.8 - 2t + 2t$$
$$12t = 4.8$$
$$\frac{\cancel{12}^{1} t}{\cancel{12}^{1}} = \frac{4.8}{12}$$
$$t = 0.4 \text{ hour}$$

The distance he rode is $D = RT$ or $D = 10 \times 0.4 = 4$ miles.

Evaluation: You can check the answer by determining the distance the person walked.

$$D = RT$$
$$D = 2(2.4 - 0.4)$$
$$= 2(2)$$
$$= 4 \text{ miles}$$

In the previous example, the same person made a round trip. In the next example, we have two vehicles going in the same direction.

EXAMPLE

A boater can travel from Port Clinton to Smithton in 3 hours. If he goes 5 miles per hour faster, he can travel the same distance in 45 minutes less. How far is it from Port Clinton to Smithton? (Ignore the current.)

SOLUTION

Goal: You are being asked the distance between the two ports.

Strategy: In this case, both trips are in the same direction. See Figure 8-2. Let R = the rate on the first trip and $R + 5$ be the rate on the second trip. Place these values in the table under Rate.

Port Clinton   Distance   Smithton

Port Clinton   Distance   Smithton

FIGURE 8-2

Place the times, 3 hours and $2\frac{1}{4}$ hours, in the time boxes. (Note $3 - \frac{3}{4} = 2\frac{1}{4}$.)

To get the distance, multiply the rate by the time for each case.

	Rate	×	Time	=	Distance
Trip 1	r		3		$3r$
Trip 2	$r + 5$		$2\frac{1}{4}$		$2\frac{1}{4}(r + 5)$

Since the distances are the same, the equation is $3r = 2\frac{1}{4}(r + 5)$.

Implementation: Solve the equation:

$$3r = 2\frac{1}{4}(r + 5)$$

$$3r = 2\frac{1}{4}r + 2\frac{1}{4} \cdot 5$$

$$3r - 2\frac{1}{4}r = 2\frac{1}{4}r - 2\frac{1}{4}r + 11\frac{1}{4}$$

$$\frac{3}{4}r = 11\frac{1}{4}$$

$$\left(\frac{3}{4}\right)r \div \frac{3}{4} = 11\frac{1}{4} \div \frac{3}{4}$$

$$r = 15 \text{ miles per hour}$$

To find the distance, use the formula $D = RT$.

$D = 15 \times 3 = 45$ miles

Hence the distance between Port Clinton and Smithton is 45 miles.

Evaluation: **Check to see if the other distance is the same.**

$$D = RT$$

$$D = 2\frac{1}{4}(15 + 5)$$

$$D = 2\frac{1}{4}(20)$$

$$= 45 \text{ miles}$$

Another type of distance problem is one where two vehicles are going in the opposite direction.

EXAMPLE

Two hikers 12.5 miles apart begin by walking toward each other, and they meet in 2.5 hours. If one hiker walks one mile farther in an hour than the other, how fast does each hiker walk?

SOLUTION

Goal: **You are being asked to find the speed in miles per hour at which each hiker walks.**

Strategy: **Draw a diagram showing that each person was walking toward the other or in the opposite directions. See Figure 8-3. Let x be the rate of the first hiker and $x + 1$ be the rate of the second hiker. Place these expressions in the boxes under Rate. The time both hikers walk is 2.5 hours. Place this value in the boxes under Time. Then under Distance, write $2.5x$ and $2.5(x + 1)$.**

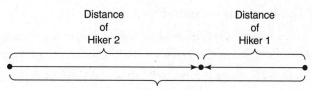

Distance of Hiker 2   Distance of Hiker 1

12.5 miles

FIGURE 8-3

	Rate	×	Time	=	Distance
Hiker 1	x		2.5		$2.5x$
Hiker 2	$x + 1$		2.5		$2.5(x + 1)$

Since the sum of the distances is 12.5 miles, the equation is $2.5x + 2.5(x + 1) = 12.5$.

Implementation: Solve the equation for x:

$$2.5x + 2.5(x + 1) = 12.5$$

$$2.5x + 2.5x + 2.5 = 12.5$$

$$5x + 2.5 = 12.5$$

$$5x + 2.5 - 2.5 = 12.5 - 2.5$$

$$5x = 10$$

$$\frac{\cancel{5}^{1} x}{\cancel{5}^{1}} = \frac{10}{5}$$

$$x = 2 \text{ miles per hour (rate of Hiker 1)}$$

$$x + 1 = 2 + 1 = 3 \text{ miles per hour (rate of Hiker 2)}$$

Evaluation: In order to check the answer, find the distances each hiker walked, then see if the sum is equal to 12.5 miles.

Hiker 1	Hiker 2
$D = RT$	$D = RT$
$D = 2 \times 2.5$	$D = 3 \times 2.5$
$= 5$ miles	$= 7.5$ miles

$$5 \text{ miles} + 7.5 \text{ miles} = 12.5 \text{ miles}$$

TRY THESE

1. A girl ran to her friend's home at the rate of 5 miles per hour and walked home at the rate of 3 miles per hour. If it took 12 minutes for the round trip, how far away is her friend's house?

2. On a trip, a motorist travels an average of 30 miles per hour in town and 60 miles per hour on the freeway. If a trip of 60 miles took him an hour and a half, how many miles did he drive on the freeway?

3. A freight train and an AMTRAK train leave towns that are 450 miles apart and travel toward each other. They pass each other in 5 hours. The AMTRAK train travels 20 miles per hour faster than the freight train. Where do they meet?

4. One person riding a motorcycle leaves 90 minutes after another person riding another motorcycle leave from the same place, riding in the same

direction. If the person riding the first motorcycle travels 30 miles per hour and the person riding the second motorcycle travels at 40 miles per hour, how long will it take the second person to overtake the first person?

5. A person riding a motorcycle leaves a city at the same time as another person driving an automobile. They travel in opposite directions. If the person riding the motorcycle is traveling 25 miles per hour and the person driving the automobile is traveling 45 miles per hour, how long will it be before they are 280 miles apart?

 SOLUTIONS

1. Minutes must be changed to hours since the rates are given in miles per hour. Twelve minutes $= \dfrac{12}{60}$ hour or 0.2 hour. Let $t =$ the time it took the girl to run to her friend's house and $0.2 - t =$ the time it took her to walk back home.

	Rate	×	Time	=	Distance
To friend's house	5		t		$5t$
Return home	3		$0.2 - t$		$3(0.2 - t)$

The distances are equal since she is making a round trip.

$$5t = 3(0.2 - t) \qquad D = 5t$$
$$5t = 3(0.2) - 3t \qquad = 5(0.075)$$
$$5t = 0.6 - 3t \qquad = 0.375 \text{ miles}$$
$$5t + 3t = 0.6 - 3t + 3t$$
$$8t = 0.6$$
$$\frac{\cancel{8}^{1}}{\cancel{8}^{1}} = \frac{0.6}{8}$$
$$t = 0.075 \text{ hour}$$

2. Let $t =$ the time the driver drove in town and $1.5 - t =$ the time the driver drove on the freeway (one hour and a half = 1.5 hours).

	Rate	×	Time	=	Distance
Town	30		t		$30t$
Freeway	60		$1.5 - t$		$60(1.5 - t)$

Since the total distance is 60 miles, the equation is $30t + 60(1.5 - t) = 60$.

$$30t + 60(1.5 - t) = 60$$

$$30t + 90 - 60t = 60$$

$$-30t + 90 = 60$$

$$-30t + 90 - 90 = 60 - 90$$

$$-30t = -30$$

$$\frac{\cancel{30}^{1}\, t}{\cancel{30}_{1}} = \frac{-30}{-30}$$

$$t = 1 \text{ hour (time on town roads)}$$

$$1.5 - t = 1.5 - 1 = 0.5 \text{ hour (time on freeway)}$$

$$D = RT$$

$$D = 0.5(60)$$

$$= 30 \text{ miles}$$

3. Let x = the rate (speed) of the freight train and $x + 20$ = the speed of the AMTRAK train.

	Rate ×	Time =	Distance
Freight train	x	5	$5x$
AMTRAK	$x + 20$	5	$5(x + 20)$

Since they meet at some point, the total distance both travel is 450 miles.

$$5x + 5(x + 20) = 450$$

$$5x + 5x + 100 = 450$$

$$10x + 100 = 450$$

$$10x + 100 - 100 = 450 - 100$$

$$10x = 350$$

$$\frac{\cancel{10}^{1}\, x}{\cancel{10}_{1}} = \frac{350}{10}$$

$$x = 35 \text{ miles per hour (speed of the freight train)}$$

$$D = RT = 35 \times 5 = 175 \text{ miles from the starting point of the}$$
freight train.

4. Let t = the time the first motorcycle travels and $t - 1.5$ be the time the second motorcycle travels. Ninety minutes $= \dfrac{90}{60} = 1.5$ hours.

	Rate	×	Time	=	Distance
First motorcycle	30		t		$30t$
Second motorcycle	40		$t - 1.5$		$40(t - 1.5)$

They travel the same distance.

$$30t = 40(t - 1.5)$$
$$30t = 40t - 60$$
$$30t - 40t = 40t - 40t - 60$$
$$-10t = -60$$
$$\frac{\cancel{-10}^{\,1}\,t}{\cancel{-10}^{\,1}} = \frac{-60}{-10}$$
$$t = 6 \text{ hours}$$
$$t - 1.5 = 6 - 1.5 = 4.5 \text{ hours}$$

The second motorcycle will overtake the first motorcycle 4.5 hours after starting.

5. Let t = the time both drivers travel.

	Rate	×	Time	=	Distance
Motorcycle	25		t		$25t$
Automobile	45		t		$45t$

The total distance that they travel is 280 miles.

$$25t + 45t = 280$$
$$70t = 280$$
$$\frac{\cancel{70}^{\,1}\,t}{\cancel{70}^{\,1}} = \frac{280}{70}$$
$$t = 4 \text{ hours}$$

Automobile	Motorcycle
$D = RT$	$D = RT$
$= 45 \cdot 4$	$= 25 \cdot 4$
$= 180$ miles	$= 100$ miles

The total distance that they travel is $100 + 180 = 280$ miles.

The motorcycle and the automobile are 280 miles apart after 4 hours.

Another type of distance problem involves an airplane flying *with* or *against* the wind or a boat moving *with* or *against* the current. If an airplane is flying with a *headwind*, the speed of the airplane is slowed down by the force of the wind. If an airplane is flying with a *tailwind*, the speed of the airplane is increased by the wind. For example, if an airplane is flying at an airspeed of 150 miles per hour and there is a 30 mile-per-hour tailwind, then the ground speed of the airplane is actually $150 + 30 = 180$ miles per hour. The airspeed is the speed of the plane as shown on its speedometer, but if you were standing on the ground, you would clock the speed at 180 miles per hour. If the plane had an airspeed of 150 miles per hour and it was flying with a headwind of 30 miles per hour, the ground speed of the airplane would be $150 - 30 = 120$ miles per hour. In order to solve these problems using algebra, the direction of the wind must be parallel to the destination of the airplane. When this situation is not true, trigonometry must be used.

In a similar situation, if a boat is moving downstream at 25 miles per hour (indicated on its speedometer) and the current is 3 miles per hour, then the actual speed of the boat is $25 + 3 = 28$ miles per hour since the current is actually pushing the boat. If the boat is going upstream against the current, then the current is pushing against the boat holding it back. In this case, the speed of the boat is $25 - 3 = 22$ miles per hour.

EXAMPLE

An airplane flies from Pittsburgh to Philadelphia in 2 hours and returns in 2.5 hours. If the wind speed is 15 miles per hour blowing from the west, find the airspeed of the plane.

SOLUTION

Goal: You are being asked to find the airspeed of the plane.

Strategy: Let x = the airspeed of the plane. Since Philadelphia is east of Pittsburgh and the wind is blowing from west to east, the ground speed from Pittsburgh to Philadelphia is $x + 15$. The ground speed from Philadelphia to Pittsburgh is $x - 15$. The times are given.

	Rate	×	Time	=	Distance
To Philadelphia	$x + 15$		2		$2(x + 15)$
To Pittsburgh	$x - 15$		2.5		$2.5(x - 15)$

Since the distances are equal, the equation is $2(x + 15) = 2.5(x - 15)$.

Implementation: Solve the equation:

$$2(x + 15) = 2.5(x - 15)$$
$$2x + 30 = 2.5x - 37.5$$
$$2x - 2x + 30 = 2.5x - 2x - 37.5$$
$$30 = 0.5x - 37.5$$
$$30 + 37.5 = 0.5x - 37.5 + 37.5$$
$$67.5 = 0.5x$$
$$\frac{67.5}{0.5} = \frac{\cancel{0.5}^{1} x}{\cancel{0.5}^{1}}$$

135 miles per hour $= x$

Evaluation: Check to see if the distances are the same. Use $D = RT$.

$$\text{To Philadelphia: } D = 2(x + 15)$$
$$= 2(135 + 15)$$
$$= 300 \text{ miles}$$
$$\text{To Pittsburgh: } D = 2.5(x - 15)$$
$$= 2.5(135 - 15)$$
$$= 2.5(120)$$
$$= 300 \text{ miles}$$

EXAMPLE

A boat's speedometer reads 22 miles per hour going downstream and reaches its destination in an hour. If the return trip takes one and a half hours at the speed of 25 miles per hour, how fast is the current?

SOLUTION

Goal: You are being asked to find the speed (rate) of the current.

Strategy: Let x = the rate of the current; then the speed of the boat downstream is $22 + x$ and upstream is $25 - x$. The times are given.

	Rate	×	Time	=	Distance
Downstream	$22 + x$		1		$1(22 + x)$
Upstream	$25 - x$		1.5		$1.5(25 - x)$

Implementation: Solve the equation:

$$1(22 + x) = 1.5(25 - x)$$

$$22 + x = 37.5 - 1.5x$$

$$22 - 22 + x = 37.5 - 22 - 1.5x$$

$$x = 15.5 - 1.5x$$

$$x + 1.5x = 15.5 - 1.5x + 1.5x$$

$$2.5x = 15.5$$

$$\frac{\cancel{2.5}^{1} x}{\cancel{2.5}^{1}} = \frac{15.5}{2.5}$$

$$x = 6.2 \text{ miles per hour}$$

Evaluation: Check to see if the distance going downstream is equal to the distance going upstream, using $D = RT$.

$$\text{Downstream } D = 1(22 + x)$$

$$= 1(22 + 6.2)$$

$$= 28.2 \text{ miles}$$

$$\text{Upstream } D = 1.5(25 - x)$$

$$= 1.5(25 - 6.2)$$

$$= 1.5(18.8)$$

$$= 28.2 \text{ miles}$$

TRY THESE

1. A plane flies with a headwind of 27 miles per hour from LeMont to Pleasantville in 5 hours and returns in 3.3 hours with a tailwind of 27 miles per hour. Find the distance between the airports.

2. A plane flies from New Eagle to South Pine in 3 hours and returns in 5 hours. If the speed of the wind is 25 miles per hour and it is blowing in the direction of South Pine from New Eagle, find the airspeed of the plane.

3. A boat's speed on its speedometer reads 12 miles per hour going downstream, and it reaches its destination in 1.6 hours. The return trip takes 3 hours at 10 miles per hour on the speedometer. Find the speed of the current.

4. If a plane flies from Unity to South Chester in 6 hours with a headwind of 24 miles per hour and returns in 4.2 hours with a tailwind of 18 miles per hour, find the airspeed of the plane.

5. If a boat travels upstream from Allentown to Bolder City in 3 hours and returns downstream from Bolder City to Allentown in 1.8 hours, find the speed of the boat (on its speedometer) if the current is 2 miles per hour.

✔ SOLUTIONS

1. Let x = the airspeed of the plane.

	Rate ×	Time =	Distance
To Pleasantville	$x - 27$	5	$5(x - 27)$
To Lemont	$x + 27$	3.3	$3.3(x + 27)$

$$5(x - 27) = 3.3(x + 27)$$

$$5x - 135 = 3.3x + 89.1$$

$$5x - 3.3x - 135 = 3.3x - 3.3 + 89.1$$

$$1.7x - 135 = 89.1$$

$$1.7x - 135 + 135 = 89.1 + 135$$

$$1.7x = 224.1$$

$$\frac{\cancel{1.7}^{1} x}{\cancel{1.7}^{1}} = \frac{224}{1.7}$$

$$x = 131.82 \text{ miles per hour (rounded)}$$

To find the distance between the airports, find $5(x - 27)$

$$5(x - 27) = 5 (131.82 - 27)$$

$$= 5 (104.85)$$

$$= 524.25$$

The airports are 524.25 miles apart.

2. Let x = the airspeed of the plane.

	Rate	×	Time	=	Distance
To South Pine	$x + 25$		3		$3(x + 25)$
To New Eagle	$x - 25$		5		$5(x - 25)$

$$3(x + 25) = 5(x - 25)$$

$$3x + 75 = 5x - 125$$

$$3x - 5x + 75 = 5x - 5x - 125$$

$$-2x + 75 = -125$$

$$-2x + 75 - 75 = -125 - 75$$

$$-2x = -200$$

$$\frac{-2x}{-2} = \frac{-200}{-2}$$

$$x = 100 \text{ miles per hour}$$

3. Let x = the speed of the current.

	Rate	×	Time	=	Distance
Downstream	$12 + x$		1.6		$1.6(12 + x)$
Upstream	$10 - x$		3		$3(10 - x)$

$$1.6(12 + x) = 3(10 - x)$$

$$19.2 + 1.6x = 30 - 3x$$

$$19.2 + 1.6x + 3x = 30 - 3x + 3x$$

$$19.2 + 4.6x = 30$$

$$19.2 - 19.2 + 4.6x = 30 - 19.2$$

$$4.6x = 10.8$$

$$\frac{4.6x}{4.6} = \frac{10.8}{4.6}$$

$$x = 2.35 \text{ miles per hour (rounded)}$$

4. Let x = the airspeed of the airplane.

	Rate	×	Time	=	Distance
To South Chester	$x - 24$		6		$6(x - 24)$
To Unity	$x + 18$		4.2		$4.2(x + 18)$

The distances are the same.

$$6(x - 24) = 4.2(x + 18)$$

$$6x - 144 = 4.2x + 75.6$$

$$6x - 4.2x - 144 = 4.2x - 4.2x + 75.6$$

$$1.8x - 144 = 75.6$$

$$1.8x - 144 + 144 = 75.6 + 144$$

$$1.8x = 219.6$$

$$\frac{\cancel{1.8}^{1} x}{\cancel{1.8}_{1}} = \frac{219.6}{1.8}$$

$$x = 122 \text{ miles per hour}$$

5. Let x = the speed of the boat.

	Rate	×	Time	=	Distance
To Bolder City	$x - 2$		3		$3(x - 2)$
To Allentown	$x + 2$		1.8		$1.8(x + 2)$

The distances are the same.

$$3(x - 2) = 1.8(x + 2)$$

$$3x - 6 = 1.8x + 3.6$$

$$3x - 1.8x - 6 = 1.8x - 1.8x + 3.6$$

$$1.2x - 6 = 3.6$$

$$1.2x - 6 + 6 = 3.6 + 6$$

$$1.2x = 9.6$$

$$\frac{\cancel{1.2}^{1} x}{\cancel{1.2}_{1}} = \frac{9.6}{1.2}$$

$$x = 8 \text{ miles per hour}$$

In this section, you learned how to solve word problems involving distance. You use the basic formula, distance = rate × time.

Mixture Problems

Many real-life problems involve mixtures. There are three basic types of mixture problems. One type uses percents. For example, a metal worker may wish to combine two alloys of different percentages of copper to make a third alloy consisting of a specific percentage of copper. In this case, it is necessary to remember that the percent of the specific substance in the mixture times the amount of mixture is equal to the amount of the pure substance in the mixture. Suppose you have 64 ounces of a mixture consisting of alcohol and water, and 30% of it is alcohol. Then 30% of 64 ounces or 19.2 ounces of the mixture is alcohol. Another type of problem involves diluting solutions. Finally, mixture problems can also include mixing nuts, candies, etc. These types of problems are explained in this section.

A table can be used to solve the percent mixture problems and an equation can be written using

$$\text{Mixture 1} + \text{Mixture 2} = \text{Mixture 3}$$

Note: The word mixture applies to alloy, solution, etc.

EXAMPLE

A metallurgist has two alloys of copper. The first one is 40% copper and the second one is 70% copper. How many ounces of each must be mixed to have 24 ounces of an alloy that is 50% copper?

SOLUTION

Goal: You are being asked to find how much of each alloy should be mixed to get 24 ounces of an alloy that is 50% copper.

Strategy: Let x = the amount of the 40% copper alloy and $24 - x$ = the amount of the 70% copper alloy; then

	Amount ×	Percent =	Total
Alloy 1	x	40%	40%(x)
Alloy 2	$24 - x$	70%	70%($24 - x$)
Alloy 3	24	50%	50%(24)

The equation is

Alloy 1	+	Alloy 2	=	Alloy 3
40%(x)	+	70%$(24-x)$	=	50%(24)

Implementation: Solve the equation:

$$40\%(x) + 70\%(24 - x) = 50\%(24)$$

Change the percents to decimals before solving the equation.

$$0.40x + 0.70(24 - x) = 0.50(24)$$

$$0.40x + 16.8 - 0.70x = 12$$

$$-0.30x + 16.8 = 12$$

$$-0.30x + 16.8 - 16.8 = 12 - 16.8$$

$$-0.30x = -4.8$$

$$\frac{-0.30\,x}{-0.30} = \frac{-4.8}{-0.30}$$

$$x = 16 \text{ ounces of Alloy 1}$$

$$24 - x = 24 - 16 = 8 \text{ ounces of Alloy 2}$$

Hence, 16 ounces of the 40% alloy should be mixed with 8 ounces of the 70% alloy to get 24 ounces of an alloy that is 50% copper.

Evaluation: Check the solution:

$$40\%(x) + 70\%(24 - x) = 50\%(24)$$

$$0.40(16) + 0.70(24 - 16) = 0.50(24)$$

$$6.4 + 5.6 = 12$$

$$12 = 12$$

EXAMPLE

A pharmacist has two bottles of alcohol; one bottle contains a 60% solution of alcohol and the other bottle contains a 85% solution of alcohol. How much of each should be mixed to get 30 ounces of a solution that is 75% alcohol?

SOLUTION

Goal: You are being asked to find the amounts of each solution that need to be mixed to get 30 ounces of a 75% alcohol solution.

Strategy: Let x = the amount of the 60% solution and $30 - x$ = the amount of the 45% solution; then set up a table as follows:

	Amount	×	Percent	=	Total
Solution 1	x		60%		60%x
Solution 2	$30 - x$		85%		85%(30 − x)
Solution 3	30		75%		75%(30)

The equation is

Solution 1	+	Solution 2	=	Solution 3
60%x	+	85%(30 – x)	=	75%(30)

Implementation: Solve the equation:

$$60\%x + 85\%(30 - x) = 75\%(30)$$

Change the percents to decimals before solving the equation.

$$0.60x + 0.85x(30 - x) = 0.75(30)$$

$$0.60x + 25.5 - 0.85x = 22.5$$

$$-0.25x + 25.5 = 22.5$$

$$-0.25x + 25.5 - 25.5 = 22.5 - 25.5$$

$$0.25x = -3$$

$$\frac{-0.25x}{-0.25} = \frac{-3}{-0.25}$$

$$x = 12 \text{ ounces of Solution 1}$$

$$30 - x = 30 - 12 = 18 \text{ ounces of Solution 2}$$

Hence, 12 ounces of the 60% solution should be mixed with 18 ounces of the 85% solution to get 30 ounces of a 75% solution.

Evaluation: Check the solution:

$$60\%x + 85\%(30 - x) = 75\%(30)$$

$$60\%(12) + 85\%(30 - 12) = 75\%(30)$$

$$0.60(12) + 0.85(18) = 0.75(30)$$

$$7.2 + 15.3 = 22.5$$

$$22.5 = 22.5$$

The second type of mixture problem involves diluting a solution or alloy. In these types of problems you are adding a weaker solution or alloy to bring down the concentration of the substance. Here you let x be the amount of the weaker solution or alloy that is being added to the original solution. Again, the equation is

$$\text{Mixture 1} + \text{Mixture 2} = \text{Mixture 3}$$

EXAMPLE _____

How much water needs to be added to 32 ounces of a 30% alcohol solution to dilute it to a 20% alcohol solution?

✓SOLUTION _____

Let x = the amount of water that needs to be added. Since there is no alcohol in pure water, the percent is 0%.

	Amount	×	Percent	=	Total
Solution 1	32		30%		30%(32)
Solution 2	x		0%		0%x
Solution 3	32 + x		20%		20%(32 + x)

Solution 1	+	Solution 2	=	Solution 3
30%(32)	+	0%x	=	20%(32 + x)

$$0.30(32) + 0x = 0.20(32 + x)$$
$$9.6 + 0 = 6.4 + 0.20x$$
$$9.6 - 6.4 = 6.4 - 6.4 + 0.2x$$
$$3.2 = 0.20x$$

$$\frac{3.2}{0.20} = \frac{\cancel{0.20}^{1}\, x}{\cancel{0.20}^{1}}$$

$$16 \text{ ounces} = x$$

Hence, 16 ounces of water must be added to the 30% solution to get a solution that is 20% alcohol.

Evaluation: **Check the solution:**

$$30\%(32) + 0\%x = 20\%(32 + x)$$
$$0.30(32) + 0 = 0.20(32 + 16)$$
$$0.30(32) = 0.20(48)$$
$$9.6 = 9.6$$

The third type of mixture problem consists of mixing two items such as coffees, teas, candy, etc., with different prices. These problems are similar to the previous ones. You can use this basic equation:

(Amount of item 1) (Its price) + (Amount of item 2) (Its price) = (Mixture amount) (Its price)

EXAMPLE

A merchant mixes some candy costing $6 a pound with some candy costing $2 a pound. How much of each must be used in order to make 25 pounds of mixture costing $4 per pound?

SOLUTION

Goal: You are being asked to find how much of each candy must be mixed together to get 25 pounds of candy costing $4.

Strategy: Let x = the amount of the $6 candy and $25 - x$ = the amount of the $2 candy; then

	Amount ×	Price =	Total
Candy 1	x	$6	$6x$
Candy 2	$25 - x$	$2	$2(25 - x)$
Mixture	25	$4	$4(25)$

The equation is $6x + 2(25 - x) = 4(25)$.

Implementation: Solve the equation:

$$6x + 2(25 - x) = 4(25)$$
$$6x + 50 - 2x = 100$$
$$4x + 50 = 100$$
$$4x + 50 - 50 = 100 - 50$$
$$4x = 50$$
$$\frac{\cancel{4}^{1} x}{\cancel{4}^{1}} = \frac{50}{4}$$
$$x = 12.5 \text{ pounds of } \$6 \text{ candy}$$
$$25 - x = 25 - 12.5 = 12.5 \text{ pounds of } \$2 \text{ candy}$$

Hence, 12.5 pounds of candy costing $6 per pound must be mixed with 12.5 pounds of candy costing $2 a pound to get 25 pounds of candy costing $4 a pound.

Evaluation: Check the solution:

$$6x + 2(25 - x) = 4(25)$$
$$6(12.5) + 2(25 - 12.5) = 4(25)$$
$$75 + 25 = 100$$
$$100 = 100$$

TRY THESE

1. A store owner wants to mix some fudge that sells for $4.50 a pound with some fudge that sells for $6 a pound. How much of each kind of fudge must he mix in order to get a 25-pound mixture that sells for $5.50 a pound?

2. How much of a solution that contains 40% alcohol must be mixed with a solution that contains 72% alcohol to get 600 milliliters of a solution that is 54% alcohol?

3. A chemist has 15% and 25% solutions of glycerol and alcohol. How much of each should be mixed to get 10 ounces of a 22% solution?

4. How many ounces of water must be added to 32 ounces of a 60% alcohol solution to dilute it to a 40% solution?

5. A grocer wants to sell some nuts for $3 a pound. How many pounds of nuts that sell for $5 a pound should be mixed with nuts that sell for $2 a pound to get a mixture of 24 pounds of nuts that sell for $3 a pound?

6. A goldsmith wants to make 50 ounces of a gold alloy that is 48% gold by mixing an alloy that contains 60% gold with one that contains 25% gold. How many ounces of each type should be mixed?

7. A candy maker wants to make 50 one-pound boxes of mixed candy that sell for $2 a box. He has on hand 20 pounds of candy that sells for $1.50 a pound. What should be the price of the other candy that he will use?

8. A baker wants to mix 10 pounds of cookies costing $2 a pound with some cookies costing $3.50 a pound. How many pounds of the $3.50 cookies should be mixed with the 10 pounds of $2 cookies to get a mixture of cookies costing $2.75 a pound?

9. A merchant wants to sell some tea costing $4 a pound. She has 15 pounds of tea costing $2.50 a pound. How many pounds of tea costing $5 per pound should she mix with 15 pounds of the $2 tea to get a mixture costing $4 a pound?

10. How much of an alloy that is 60% zinc should be added to 120 pounds of an alloy that is 40% zinc to get an alloy that is 54% zinc?

✔SOLUTIONS

1. Let x = the amount of fudge that sells for $4.50 a pound and $(25 - x)$ = the amount of fudge that sells for $6 a pound.

	Amount	×	Price	=	Total
Fudge 1	x		$4.50		$4.5x$
Fudge 2	$25 - x$		$6		$6(25 - x)$
Mixture	25		$5.50		5.5(25)

Mixture 1	+	Mixture 2	=	Mixture 3
$4.50x	+	$6(25 - x)	=	$5.50(25)

$$4.5x + 6(25 - x) = 5.5(25)$$

$$4.5x + 150 - 6x = 137.5$$

$$-1.5x + 150 = 137.5$$

$$-1.5x + 150 - 150 = 137.5 - 150$$

$$-1.5x = -12.5$$

$$\frac{\cancel{1.5}^{1}\, x}{\cancel{1.5}^{1}} = \frac{-12.5}{-1.5}$$

$$x = 8\frac{1}{3} \text{ pounds or } 8.33 \text{ pounds (rounded)}$$

$$25 - x = 25 - 8\frac{1}{3} = 16\frac{2}{3} \text{ pounds or } 16.67 \text{ pounds (rounded)}$$

Hence, to get the proper mixture, the store owner should mix $8\frac{1}{3}$ pounds of the $4.50 fudge with $16\frac{2}{3}$ pounds of the $6 fudge.

2. Let x = the amount of the 40% solution and $(600 - x)$ = the amount of the 72% solution.

	Amount	×	Percent	=	Total
Solution 1	x		40%		40%(x)
Solution 2	$600 - x$		72%		72%(600 − x)
Solution 3	600		54%		54%(600)

Solution 1	+	Solution 2	=	Solution 3
40%(x)	+	72%(600 − x)	=	54%(600)

$$0.40x + 0.72(600 - x) = 0.54(600)$$

$$0.40x + 432 - 0.72x = 324$$

$$-0.32x + 432 = 324$$

$$-0.32x + 432 - 432 = 324 - 432$$

$$-0.32x = -108$$

$$\frac{\cancel{-0.32}^{1}x}{\cancel{-0.32}_{1}} = \frac{-108}{-0.32}$$

$$x = 337.5 \text{ milliliters}$$

$$600 - x = 600 - 337.5 = 262.5 \text{ milliliters}$$

Hence, 337.5 milliliters of the 40% solution must be mixed with 262.5 milliliters to get 600 milliliters of a 54% solution.

3. Let x = the amount of the 15% solution and $(10 - x)$ = the amount of the 25% solution.

	Amount	×	Percent	=	Total
Solution 1	x		15%		15%(x)
Solution 2	$(10 - x)$		25%		25%(10 − x)
Solution 3	10		22%		22%(10)

Solution 1	+	Solution 2	=	Solution 3
15%(x)	+	25%(10 − x)	=	22%(10)

$$0.15x + 0.25(10 - x) = 0.22(10)$$

$$0.15x + 2.5 - 0.25x = 2.2$$

$$-0.10x + 2.5 = 2.2$$

$$-0.10x + 2.5 - 2.5 = 2.2 - 2.5$$

$$-0.10x = -0.3$$

$$\frac{\cancel{-0.10}^{\,1}\,x}{\cancel{-0.10}_{\,1}} = \frac{-0.3}{-0.10}$$

$$x = 3 \text{ ounces}$$

$$10 - x = 10 - 3 = 7 \text{ ounces}$$

Hence, the chemist would have to mix 3 ounces of the 15% solution and 7 ounces of the 25% solution to get 10 ounces of the 22% solution.

4. Let $x =$ the amount of water to be added to the solution. A 60% solution of alcohol is 40% water (100% – 60%).

	Amount	×	Percent	=	Total
Solution 1	32		60%		60%(32)
Solution 2	x		0%		0%(x)
Solution 3	32 + x		40%		40%(32 + x)

Solution 1	+	Solution 2	=	Solution 3
60%(32)	+	0%(x)	=	40%(32 + x)

$$0.60(32) + 0 = 0.40(32 + x)$$

$$0.60(32) + 0 = 12.8 + 0.40x$$

$$19.2 = 12.8 + 0.40x$$

$$19.2 - 12.8 = 12.8 - 12.8 + 0.40x$$

$$6.4 = 0.40x$$

$$\frac{6.4}{0.40} = \frac{\cancel{0.40}^{\,1}\,x}{\cancel{0.40}_{\,1}}$$

$$16 \text{ ounces} = x$$

Hence, if 16 ounces of water is added to a solution that is 60% alcohol, it will dilute it to a solution that is 40% alcohol.

5. Let $x =$ the amount of nuts that sell for $5 a pound and $24 - x =$ the amount of nuts that sell for $2 a pound.

	Amount	×	Price	=	Total
Mixture 1	x		$5		5x$
Mixture 2	24 – x		$2		$2(24 – x)
Mixture 3	24		$3		$3(24)

Mixture 1	+	Mixture 2	=	Mixture 3
$5x	+	$2(24 – x)	=	$3(24)

$$5x + 2(24 - x) = 3(24)$$

$$5x + 48 - 2x = 72$$

$$3x + 48 = 72$$

$$3x + 48 - 48 = 72 - 48$$

$$3x = 24$$

$$\frac{\cancel{3}^{1} x}{\cancel{3}_{1}} = \frac{24}{3}$$

$$x = 8$$

$$24 - x = 24 - 8 = 16 \text{ pounds}$$

Hence, the grocer should mix 8 pounds of the $5 mix and 16 pounds of the $2 mix to get 24 pounds of mixed nuts that sell for $3 a pound.

6. Let x = the amount of the alloy that is 60% gold and $50 - x$ = the amount of the alloy that is 25% gold.

	Amount	×	Percent	=	Total
Mixture 1	x		60%		60%(x)
Mixture 2	$50 - x$		25%		25%(50 – x)
Mixture 3	50		48%		48%(50)

Mixture 1	+	Mixture 2	=	Mixture 3
60%x	+	25%(50 – x)	=	48%(50)

$$0.60x + 0.25(50 - x) = 0.48(50)$$

$$0.60x + 12.5 - 0.25x = 24$$

$$0.35x + 12.5 = 24$$

$$0.35x + 12.5 - 12.5 = 24 - 12.5$$

$$0.35x = 11.5$$

$$\frac{\cancel{0.35}^{1} x}{\cancel{0.35}_{1}} = \frac{11.5}{0.35}$$

$$x = 32\frac{6}{7} \text{ ounces}$$

$$50 - x = 50 - 32\frac{6}{7} = 17\frac{1}{7} \text{ ounces}$$

Hence, the goldsmith should mix $32\frac{6}{7}$ ounces of the 60% alloy with $17\frac{1}{7}$ ounces of the 25% alloy to get 50 ounces of a 48% gold alloy.

7. Let x = the price of the mixture that he will use. Since he has 20 pounds, he will need 30 pounds of the other mixture (50 − 20 = 30).

	Amount	×	Price	=	Total
Mixture 1	20		$1.50		$1.50(20)
Mixture 2	30		x		x(30)
Mixture 3	50		$2		$2(50)

Mixture 1	+	Mixture 2	=	Mixture 3
$1.50(20)	+	x(30)	=	$2(50)

$$\$1.50(20) + x(30) = \$2(50)$$
$$30 + 30x = 100$$
$$30 - 30 + 30x = 100 - 30$$
$$30x = 70$$
$$\frac{\cancel{30}^{1} x}{\cancel{30}_{1}} = \frac{70}{30}$$
$$x = \$2.33 \text{ (rounded)}$$

Hence, he will need 30 pounds of a mixture that costs $2.33 a pound.

8. Let x = the amount of the $3.50 cookies.

	Amount	×	Price	=	Total
Mixture 1	10		$2		$2(10)
Mixture 2	x		$3.50		3.50x$
Mixture 3	10 + x		$2.75		$2.75(10 + x)

Mixture 1	+	Mixture 2	=	Mixture 3
$2(10)	+	3.50x$	=	$2.75(10 + x)

$$\$2(10) + \$3.50x = \$2.75(10 + x)$$
$$20 + 3.5x = 27.5 + 2.75x$$
$$20 - 20 + 3.5x = 27.5 - 20 + 2.75x$$
$$3.5x = 7.5 + 2.75x$$
$$3.5x - 2.75x = 7.5 + 2.75x - 2.75x$$

$$0.75 = 7.5$$

$$\frac{\cancel{0.75}^{1}\, x}{\cancel{0.75}^{1}} = \frac{7.5}{0.75}$$

$$x = 10$$

Hence, the baker should add 10 pounds of cookies that cost $3.50.

9. Let x = the amount of tea costing $5 a pound.

	Amount	×	Price	=	Total
Mixture 1	15		$2.50		$2.50(15)
Mixture 2	x		$5		$5x$
Mixture 3	15 + x		$4		$4(15 + x)$

Mixture 1	+	Mixture 2	=	Mixture 3
$2.50(15)	+	5x$	=	$4(15 + x)

$$2.50(15) + 5x = 4(15 + x)$$

$$37.5 + 5x = 60 + 4x$$

$$37.5 - 37.5 + 5x = 60 - 37.5 + 4x$$

$$5x = 22.5 + 4x$$

$$5x - 4x = 22.5 + 4x - 4x$$

$$1x = 22.5$$

$$x = 22.5$$

Hence, she must mix 22.5 pounds of tea costing $5.

10. Let x = the amount of the 60% zinc alloy that is to be added.

	Amount	×	Percent	=	Total
Mixture 1	x		60%		60%(x)
Mixture 2	120		40%		40%(120)
Mixture 3	120 + x		54%		54%(120 + x)

Mixture 1	+	Mixture 2	=	Mixture 3
60%x	+	40%(120)	=	54%(120 + x)

$$0.60x + 0.40(120) = 0.54(120 + x)$$

$$0.60x + 48 = 64.8 + 0.54x$$

$$0.60x - 0.54x + 48 = 64.8 + 0.54x - 0.54x$$

$$0.06x + 48 = 64.8$$

$$0.06x + 48 - 48 = 64.8 - 48$$

$$0.06x = 16.8$$

$$\frac{\cancel{0.06}^{1} x}{\cancel{0.06}^{1}} = \frac{16.8}{0.06}$$

$$x = 280 \text{ pounds}$$

Hence, 280 pounds of 60% alloy should be added.

Mixture problems can be solved by using the basic equation: Mixture 1 + Mixture 2 = Mixture 3. Mixture problems can also include types of problems where a strong mixture must be diluted to make a weaker one.

Summary

This chapter explained how to solve distance and mixture problems.

QUIZ

1. A boat travels downstream to a park in three hours and returns to its dock in five hours. If the current is 6 miles per hour, find the speed of the boat on its speedometer.

   A. 24 miles per hour

   B. 18 miles per hour

   C. 26 miles per hour

   D. 20 miles per hour

2. Evelyn and Jill leave their office at the same time and travel in opposite directions. If Jill drives 8 miles per hour faster than Evelyn, they will be 184 miles apart after two hours. How fast was Jill driving?

   A. 40 miles per hour

   B. 42 miles per hour

   C. 48 miles per hour

   D. 50 miles per hour

3. Mary leaves for a trip driving at 52 miles per hour. One-half hour later, Beth leaves on the same interstate highway traveling 62 miles per hour. How many miles will Beth have to drive before she overtakes Mary?

   A. 138.6 miles

   B. 161.2 miles

   C. 147.4 miles

   D. 153.8 miles

4. Mike leaves his house for Bentleyville, which is 200 miles away. After 3 hours, he stops for lunch for 30 minutes; then he drives 10 miles slower for the rest of the trip. If the trip takes 5 hours, what was his beginning speed?

   A. 41.5 miles per hour

   B. 48.5 miles per hour

   C. 47.8 miles per hour

   D. 63.5 miles per hour

5. Bob bikes on a trail at an average speed of 12 miles per hour. His friend Ruth bikes at an average speed of 10 miles per hour. If they start from opposite ends of a 33-mile trail, how far from Bob's starting place will they meet?

   A. 20 miles

   B. 18 miles

   C. 22 miles

   D. 15 miles

6. A chemist wants to make a 30-ounce solution of alcohol and water that is 48% alcohol. How much of a 30% alcohol solution should be mixed with a 60% alcohol solution?

   A. 12 ounces
   B. 8 ounces
   C. 15 ounces
   D. 10 ounces

7. A hardware store owner wants to mix some nails costing $4 a pound with some nails costing $2.50 a pound to get 30 pounds of nails costing $3 a pound. How many pounds of $4 nails will he use?

   A. 20 pounds
   B. 16 pounds
   C. 10 pounds
   D. 8 pounds

8. How much milk that contains 5% butterfat must be mixed with milk containing 15% butterfat to get 100 gallons of milk that is 9% butterfat?

   A. 40 gallons
   B. 32 gallons
   C. 54 gallons
   D. 60 gallons

9. A floral shop manager wants to make 10 bouquets of roses and daisies to sell for $18 a bouquet. If the roses sell for $25 a bouquet and the daisies sell for $15 a bouquet, how many bouquets of roses will the manager need?

   A. 12 bouquets
   B. 15 bouquets
   C. 3 bouquets
   D. 8 bouquets

10. How many quarts of an iced tea drink that sells for $2 a quart must be mixed with a lemonade drink that sells for $1.20 a quart to get 12 quarts of lemonade/iced tea drink that will sell for $1.50 a quart?

   A. 5 quarts
   B. 6 quarts
   C. 4.5 quarts
   D. 8.5 quarts

chapter 9

Solving Finance, Lever, and Work Problems

This chapter explains how to solve finance problems, lever problems, and work problems.

Finance problems involve investing money at specific interest rates and receiving the interest from these investments.

Lever problems involve placing people or weights on a board that sits on a fulcrum in order to balance the board. If the weights are different from each other, they can be placed at various distances from the fulcrum in order to balance the lever. A common use of the lever is a child's seesaw.

Work problems involve two or more people performing a job. Each person works at a different rate. When the people work together, the job will take less time than the workers doing the entire job alone. These problems could also include two pipes filling or draining a tank at the same time.

CHAPTER OBJECTIVES

In this chapter, you will learn how to

- Solve finance problems
- Solve lever problems
- Solve work problems

Finance Problems

Finance problems use the basic concepts of investment. There are three terms that are used. The *interest*, also called the *return*, is the amount of money that is made on an investment. The *principal* is the amount of money invested, and the *rate* or *interest rate* is a percent that is used to compute the interest. The basic formula is

$$\text{Interest} = \text{Principal} \times \text{Rate} \times \text{Time or } I = PRT.$$

In these problems, the interest used is called simple interest, and it is the interest for one year. The problems can be set up using a table similar to the ones used in the previous lessons. The equation is derived from the following:

Interest from first investment + Interest from second investment = Total interest.

Note: Interest rates vary from time to time; however, it doesn't matter what the rates are, the problems are done in the same way. In order to make the material understandable, rates between 2% and 10% have been used. It is the procedure that is important, not the numbers.

EXAMPLE

A person has $8,000 to invest and decides to invest part of it at 6% and the rest of it at $4\frac{1}{2}$%. If the total interest for the year from the amounts invested is $435, how much does the person have invested at each rate?

SOLUTION

Goal: You are being asked to find the amounts of money invested at 6% and $4\frac{1}{2}$%.

Strategy: Let x = the amount of money invested at 6% and ($8,000 – x$) = the amount of money invested at $4\frac{1}{2}$%. Then set up a table as shown.

	Principal ×	Rate =	Interest
First investment	x	6%	6%x
Second investment	8,000 – x	$4\frac{1}{2}$%	$4\frac{1}{2}$%(8,000 – x)

The equation is

Interest on the first investment + Interest on second investment = Total interest

$$6\%x \quad + \quad 4\tfrac{1}{2}\%(8,000 - x) \quad = \$435$$

Implementation: Solve the equation:

$$6\%x + 4\tfrac{1}{2}\%(8,000 - x) = 435$$

$$0.06x + 0.045(8,000 - x) = 435$$

$$0.06x + 360 - 0.045 = 435$$

$$0.015x + 360 - 360 = 435 - 360$$

$$0.015x = 75$$

$$\frac{\cancel{0.015}^{1}x}{\cancel{0.015}^{1}} = \frac{75}{0.015}$$

$$x = \$5,000 \text{ invested at } 6\%$$

$$\$8,000 - x = \$8,000 - \$5,000 = \$3,000 \text{ invested at } 4\tfrac{1}{2}\%$$

Evaluation: Find the interest on both investments separately and then add them to see if they equal $435. Use $I = PRT$ where $T = 1$.

$$\text{First investment: } I = \$5,000(6\%) = \$300$$

$$\text{Second investment: } I = \$3,000(4\tfrac{1}{2}\%) = \$135$$

$$\$300 + \$135 = \$435$$

EXAMPLE

A person has three times as much money invested at 8% as he has at 3%. If the total annual interest from the investments is $540, how much does he have invested at each rate?

✔ SOLUTION

Goal: You are being asked to find how much money is invested at 8% and 3%.

Strategy: Let x = the amount of money invested at 3% and $3x$ = the amount of money invested at 8%; then

	Principal	×	Rate	=	Interest
First investment	$3x$		8%		8%($3x$)
Second investment	x		3%		3%(x)

The equation is 8%($3x$) + 3%(x) = 540.

Implementation: Solve the equation:

$$8\%(3x) + 3\%(x) = 540$$

$$0.08(3x) + 0.03(x) = 540$$

$$0.24x + 0.03x = 540$$

$$0.27 = 540$$

$$\frac{\cancel{0.27}^{1} x}{\cancel{0.27}^{1}} = \frac{540}{0.27}$$

$$x = 2{,}000$$

$$3x = 3(2{,}000) = 6{,}000$$

Hence, the person has $2,000 invested at 3% and $6,000 invested at 8%.

Evaluation: Find the interest earned on each investment, then add, and see if the sum is $540. Use $I = PRT$ where $T = 1$.

$$\text{First investment: } I = 3\%(\$2{,}000) = \$60$$

$$\text{Second investment: } I = 8\%(6{,}000) = \$480$$

$$\$60 + \$480 = \$540$$

EXAMPLE

An investor has $600 more invested in stocks paying 9% than she has invested in bonds paying 3%. If the total interest is $162, find the amount of money invested in each.

SOLUTION

Goal: You are being asked to find the amount of each investment.

Strategy: Let x = the amount invested in bonds and $x + 600$ = the amount invested in stocks.

	Principal	×	Rate	=	Interest
Bonds	x		3%		3%(x)
Stocks	$x + 600$		9%		9%($x + 600$)

The equation is 3%x + 9%(x + 600) = $162.

Implementation: Solve the equation:

$$3\%x + 9\%(x + 600) = \$162$$

$$0.03x + 0.09(x + 600) = 162$$

$$0.03x + 0.09x + 54 = 162$$

$$0.12x + 54 - 54 = 162 - 54$$

$$0.12x = 108$$

$$\frac{\cancel{0.12}^{1}x}{\cancel{0.12}^{1}} = \frac{108}{0.12}$$

$$x = \$900$$

$$x + 600 = 900 + 600 = \$1,500$$

Hence, the person has $900 invested in bonds and $1,500 invested in stocks.

Evaluation: Find the interest for both investments and then add to see if the answer is $162. Use $I = PRT$ where $T = 1$.

$$\text{Bonds: } I = 3\%(900) = \$27$$

$$\text{Stocks: } I = 9\%(1,500) = \$135$$

$$\$27 + \$135 = \$162$$

EXAMPLE

An investor has twice as much money invested at 7% as he has invested at 3% and $400 more invested at 2% than he has invested at 3%. If the total interest from the three investments is $84, find the amounts he has invested at each rate.

SOLUTION

Goal: You are being asked to find the amounts of the three investments.

Strategy: Let x = the amount invested at 3%, $2x$ = the amount of money invested at 7%, and $x + 400$ = the amount of money invested at 2%.

	Principal	×	Rate	=	Interest
First investment	x		3%		3%x
Second investment	$2x$		7%		7%($2x$)
Third investment	$x + 400$		2%		2%($x + 400$)

The equation is 3%x + 7%($2x$) + 2%($x + 400$) = $84.

Implementation: Solve the equation:

$$3\%x + 7\%(2x) + 2\%(x + 400) = \$84$$

$$0.03x + 0.07(2x) + 0.02(x + 400) = 84$$

$$0.03x + 0.14x + 0.02x + 8 = 84$$

$$0.19x + 8 = 84$$

$$0.19x + 8 - 8 = 84 - 8$$

$$0.19x = 76$$

$$\frac{\cancel{0.19}^{1}x}{\cancel{0.19}^{1}} = \frac{76}{0.19}$$

$$x = 400$$

$$2x = 2(400) = 800$$

$$x + 400 = 400 + 400 = 800$$

Hence, the investor invested $400 at 3%, $800 at 7%, and $800 at 2%.

Evaluation: Find the three interest amounts, and add to see if you get $84. Use $I = PRT$ where $T = 1$.

$$\text{First investment: } I = 3\%(400) = \$12$$

$$\text{Second investment: } I = 7\%(800) = \$56$$

$$\text{Third investment: } I = 2\%(800) = \$16$$

$$\$12 + \$56 + \$16 = \$84$$

TRY THESE

1. An individual invested $7,000, part at 6% and the rest at 3.5%. If the total interest he earned after one year was $357.50, find the amount of each investment.

2. An individual invested a certain amount of money in a savings account paying 2% and $1,800 more than that amount in a one-year CD paying 1.5%. If the total interest for the two investments was $51.50, find the amount of money she invested in each.

3. A person invested six times as much money at $7\frac{1}{2}$% as she did at $3\frac{1}{4}$%. If the total interest from the investments at the end of the year was $193, how much did she invest at each rate?

4. An investor made two investments, one paying 9% and one paying 4%. If the total amount invested was $15,000 and the total interest she earned after one year was $800, find the amount of each investment.

5. An investor has $1,500 less invested at 6% than he has invested at 8%. If the total yearly interest from the investments is $190, find the amounts he has invested at each rate.

6. An individual invested twice as much in bonds paying 2% as he did in stocks paying 6%. If the interest at the end of the year was $468, find the amount of money he invested in each.

7. A person invested a certain amount of money in an account paying 5%. He invests five times that amount into another account paying $3\frac{1}{2}$%, and he invests $700 more than the amount in the 5% account into a third account paying 8%. If the total yearly interest from all three accounts was $5,851, find the amount he invested in each account.

8. A person has $5,000 invested at 5%. How much should be invested at 3% to have an income (yearly) interest of $1,222?

9. Ms. Smith invested some money at 6% and some money at 9%. If the yearly interest on both investments is the same and the total amount of the investments is $15,000, find the amount of each investment.

10. An investor has three investments. He has twice the amount of money invested at $5\frac{1}{2}$% as he has invested at 1% and $600 more invested at 2% as he has at 1%. If the yearly interest is $852, find the amount of each investment.

SOLUTIONS

1. Let x = the amount of money invested at 6% and $7,000 - x$ = the amount of money invested at 3.5%.

$$6\%(x) + 3.5\%(7,000 - x) = \$357.50$$
$$0.06x + 0.035(7,000 - x) = 357.5$$
$$0.06x + 245 - 0.035x = 357.5$$
$$0.025x + 245 = 357.5$$
$$0.025x + 245 - 245 = 357.5 - 245$$
$$0.025x = 112.5$$
$$\frac{\cancel{0.025}^{1}x}{\cancel{0.025}_{1}} = \frac{112.5}{0.025}$$
$$x = \$4,500 \text{ at } 6\%$$
$$7000 - x = 7,000 - 4,500 = \$2,500 \text{ at } 3.5\%$$

2. Let x = the amount of money invested at 2% and $x + \$1,800$ = the amount of money invested at 1.5%.

$$2\%(x) + 1.5\%(x + \$1,800) = 51.50$$
$$0.02x + 0.015(x + 1,800) = 51.5$$
$$0.02x + 0.015x + 27 = 51.5$$
$$0.035x + 27 = 51.5$$
$$0.035x + 27 - 27 = 51.5 - 27$$
$$0.035x = 24.5$$
$$\frac{\cancel{0.035}^{1}x}{\cancel{0.035}_{1}} = \frac{24.5}{0.035}$$
$$x = \$700 \text{ at } 2\%$$
$$x + 1,800 = 700 + 1,800 = \$2,500 \text{ at } 1.5\%$$

3. Let x = the amount of money invested at $3\frac{1}{4}\%$ and $6x$ = the amount of money invested at $7\frac{1}{2}\%$.

$$3\frac{1}{4}\%(x) + 7\frac{1}{2}\%(6x) = \$193$$
$$0.0325x + 0.075(6x) = 193$$
$$0.0325x + 0.45x = 193$$
$$0.4825x = 193$$

$$\frac{\cancel{0.4825}^{-1}x}{\cancel{0.4825}^{-1}} = \frac{193}{0.4825}$$

$$x = \$400 \text{ at } 3\frac{1}{4}\%$$

$$6x = 6(400) = \$2,400 \text{ at } 7\frac{1}{2}\%$$

4. Let x = the amount of money invested at 9% and ($15,000 – x) = the amount of money invested at 4%.

$$9\%(x) + 4\%(\$15,000 - x) = \$800$$

$$0.09x + 0.04(15,000 - x) = 800$$

$$0.09x + 600 - 0.04x = 800$$

$$0.05x + 600 = 800$$

$$0.05x + 600 - 600 = 800 - 600$$

$$0.05x = 200$$

$$\frac{\cancel{0.05}^{-1}x}{\cancel{0.05}^{-1}} = \frac{200}{0.05}$$

$$x = \$4,000 \text{ at } 9\%$$

$$\$15,000 - x = \$15,000 - \$4,000 = \$11,000 \text{ at } 4\%$$

5. Let x = the amount of money invested at 8% and x – $1,500 = the amount invested at 6%.

$$8\%(x) + 6\%(x - \$1,500) = \$190$$

$$0.08x + 0.06(x - 1,500) = 190$$

$$0.08x + 0.06x - 90 = 190$$

$$0.14x - 90 = 190$$

$$0.14x - 90 + 90 = 190 + 90$$

$$0.14x = 280$$

$$\frac{\cancel{0.14}^{-1}x}{\cancel{0.14}^{-1}} = \frac{280}{0.14}$$

$$x = \$2,000 \text{ invested at } 8\%$$

$$x - 1,500 = 2,000 - 1,500 = \$500 \text{ invested at } 6\%$$

6. Let x = the amount invested in stocks and $2x$ = the amount of money invested in bonds.

$$6\%(x) + 2\%(2x) = \$468$$
$$0.06x + 0.02(2x) = 468$$
$$0.06x + 0.04x = 468$$
$$0.10x = 468$$
$$\frac{\cancel{0.14}^{1} x}{\cancel{0.14}_{1}} = \frac{468}{0.10}$$
$$x = \$4,680 \text{ invested in stocks}$$
$$2x = 2(\$4,680) = \$9,360 \text{ invested in bonds}$$

7. Let x = the amount of money invested at 5%, $5x$ = the amount of money invested at $3\frac{1}{2}\%$, and $x + \$700$ = the amount of money invested at 8%.

$$5\%(x) + 3.5\%(5x) + 8\%(x + 700) = 5,851$$
$$0.05x + 0.035(5x) + 0.08(x + 700) = 5,851$$
$$0.05x + 0.175x + 0.08x + 56 = 5,851$$
$$0.305x + 56 - 56 = 5,851 - 56$$
$$0.305x = 5,795$$
$$\frac{\cancel{0.305}^{1} x}{\cancel{0.305}_{1}} = \frac{5,795}{0.305}$$
$$x = \$19,000 \text{ invested at 5\%}$$
$$5x = 5(19,000) = \$95,000 \text{ invested at } 3\frac{1}{2}\%$$
$$x + 700 = 19,000 + 700 = \$19,700 \text{ invested at 8\%.}$$

8. Let x = the amount of money the person should invest at 3%.

$$5\%(\$5,000) + 3\%x = \$1,222$$
$$0.05(5,000) + 0.03x = 1,222$$
$$250 + 0.03x = 1,222$$
$$250 - 250 + 0.03x = 1,222 - 250$$
$$0.03x = 972$$
$$\frac{\cancel{0.03}^{1} x}{\cancel{0.03}_{1}} = \frac{972}{0.03}$$
$$x = \$32,400 \text{ should be invested at 3\%.}$$

9. Let x = the amount of money invested at 6% and $15,000 − x = the amount of money invested at 9%. Since the interest earned on both investments is the same, the equation is 6%(x) = 9%($15,000 − x).

$$6\%(x) = 9\%(15,000 - x)$$

$$0.06x = 0.09(15,000 - x)$$

$$0.06 = 1,350 - 0.09x$$

$$0.06x + 0.09x = 1,350 - 0.09x + 0.09x$$

$$0.15x = 1,350$$

$$\frac{\cancel{0.15}^{1}x}{\cancel{0.15}^{1}} = \frac{1,350}{0.15}$$

$$x = \$9,000 \text{ at } 6\%$$

$$(15,000 - x) = 15,000 - 9,000 = \$6,000 \text{ at } 9\%$$

10. Let x = the amount of money invested at 1%, $2x$ = the amount of money invested at $5\frac{1}{2}$%, and x + $600 = the amount of money invested at 2%.

$$1\%(x) + 5\frac{1}{2}\%(2x) + 2\%(x + \$600) = \$852$$

$$0.01x + 0.055(2x) + 0.02(x + 600) = 852$$

$$0.01x + 0.11 + 0.02 + 12 = 852$$

$$0.14x + 12 = 852$$

$$0.14x + 12 - 12 = 852 - 12$$

$$0.14x = 840$$

$$\frac{\cancel{0.14}^{1}x}{\cancel{0.14}^{1}} = \frac{840}{0.14}$$

$$x = \$6,000 \text{ at } 1\%$$

$$2x - 2(6,000) = \$12,000 \text{ at } 5\frac{1}{2}\%$$

$$x + 600 = 6,000 + 600 = \$6,600 \text{ invested at } 2\%$$

In this section, you have learned how to solve finance problems. The basic formula that is used is Interest = Principal × Rate × Time or $I = PRT$. Since the interest is yearly, the time = 1 year. You get the basic equation for the problem by using

Interest from first investment + Interest from second investment = Total interest.

There are several different types of problems, so the equation can differ somewhat from the basic one.

Lever Problems

One of the oldest machines known to humans is the lever. The principles of the lever are studied in physics. Most people are familiar with the simplest kind of lever, known as the seesaw or teeterboard, often seen in parks.

The lever is a board placed on a *fulcrum* or point of support. On a seesaw, the fulcrum is in the center of the board. A child sits at either end of the board. If one child is heavier than the other child, he or she can sit closer to the center in order to balance the seesaw. This is the basic principle of the lever.

In general, the weights are placed on the ends of the board, and the distance the weight is from the fulcrum is called the *length* or *arm*. The basic principle of the lever is that the weight times the length of the arm on the left side of the lever is equal to the weight times the length of the arm on the right side of the lever, or $WL = wl$. See Figure 9-1.

Given any of the three variables, you can set up an equation and solve for the fourth one. Unless otherwise specified, assume the fulcrum is in the center of the lever.

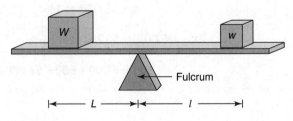

FIGURE 9-1

EXAMPLE

Sam weighs 150 pounds and sits on a seesaw 2 feet from the fulcrum. Where must Sally, who weighs 120 pounds, sit to balance it?

SOLUTION

Goal: You are being asked to find the distance from the fulcrum Sally needs to sit to balance the seesaw.

Strategy: Use the formula $WL = wl$ where $W = 150$, $L = 2$, $w = 120$, and let $x = l$.

$$WL = wl$$
$$150(2) = 120x$$

See Figure 9-2.

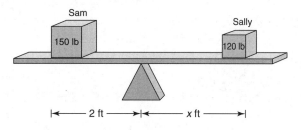

Sam   Sally

150 lb   120 lb

|← 2 ft →|← x ft →|

FIGURE 9-2

Implementation: Solve the equation:

$$150(2) = 120x$$
$$300 = 120x$$
$$\frac{300}{120} = \frac{\cancel{120}^{1} x}{\cancel{120}_{1}}$$
$$2.5 \text{ feet} = x$$

Hence, she must sit 2.5 feet from the fulcrum.

Evaluation: Check the solution:

$$WL = wl$$
$$150(2) = 120(2.5)$$
$$300 = 300$$

The fulcrum of a lever does not have to be at its center, as shown in the next example.

EXAMPLE

The fulcrum of a lever is 4 feet from the end of a 10-foot lever. On the short end rests a 96-pound weight. How much weight must be placed on the other end to balance the lever?

SOLUTION

Goal: You are being asked to find how much weight is needed to balance the lever.

Strategy: Let x = the weight of the object needed. This weight must be placed at $10 - 4 = 6$ feet from the fulcrum since it is at the end of the longer side.

$$WL = wl$$

$$96(4) = x(6)$$

See Figure 9-3.

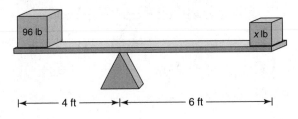

$|\!\!\leftarrow\!\!-\!\!-\!\! 4\ ft\ -\!\!-\!\!\rightarrow\!\!|\!\!\leftarrow\!\!-\!\!-\!\! 6\ ft\ -\!\!-\!\!-\!\!\rightarrow\!\!|$

FIGURE 9-3

Implementation: Solve the equation:

$$96(4) = x(6)$$

$$384 = 6x$$

$$\frac{384}{6} = \frac{\overset{1}{\cancel{6}}x}{\cancel{6}^{1}}$$

$$64 = x$$

64 pounds needs to be placed at the 6-foot end to balance the lever.

Evaluation: Check the solution:

$$WL = wl$$

$$96(4) = 64(6)$$

$$384 = 384$$

EXAMPLE

Where should the fulcrum be placed on a 12-foot lever with a 40-pound weight on one end and a 60-pound weight on the other end?

SOLUTION

Goal: You are being asked to find the placement of the fulcrum so that the lever is balanced.

Strategy: Let x = the length of the lever from the fulcrum to the 40-pound weight and $(12 - x)$ = the length of the lever from the fulcrum to the 60-pound weight. See Figure 9-4.

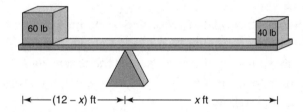

FIGURE 9-4

The equation is

$$WL = wl$$
$$40x = 60(12 - x)$$

Implementation: Solve the equation:

$$40x = 60(12 - x)$$
$$40x = 720 - 60x$$
$$40x + 60x = 720 - 60x + 60x$$
$$100x = 720$$
$$\frac{\cancel{100}^{1}x}{\cancel{100}^{1}} = \frac{720}{100}$$
$$x = 7.2$$

Hence, the fulcrum must be placed 7.2 feet from the 40-pound weight.

Evaluation: Check the solution:

$$WL = wl$$
$$40(7.2) = 60(12 - 7.2)$$
$$40(7.2) = 60(4.8)$$
$$288 = 288$$

You can place three or more weights on a lever and it still can be balanced. If four weights are used, two on each side, the equation is

$$W_1L_1 + W_2L_2 = w_1l_1 + w_2l_2$$

EXAMPLE

On a 15-foot seesaw, Mary, weighing 95 pounds, sits on one end. Next to Mary sits Helen, weighing 85 pounds. Helen is five feet from the fulcrum which is in the center of the seesaw. On the other side at the end sits Carol, weighing 105 pounds. Where should Julie, weighing 80 pounds, sit in order to balance the seesaw?

SOLUTION

Goal: You are being asked to find the distance from the fulcrum where Julie should sit in order to balance the seesaw.

Strategy: Let x = the distance from the fulcrum where Julie needs to sit. See Figure 9-5.

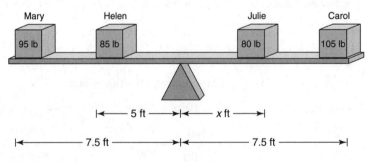

FIGURE 9-5

The equation is

$$W_1L_1 + W_2L_2 = w_1l_1 + w_2l_2$$
$$95(7.5) + 85(5) = 80x + 105(7.5)$$

Implementation: Solve the equation:

$$95(7.5) + 85(5) = 80x + 105(7.5)$$
$$712.5 + 425 = 80x + 787.5$$
$$1,137.5 = 80x + 787.5$$
$$1,137.5 - 787.5 = 80x + 787.5 - 787.5$$
$$350 = 80x$$
$$\frac{350}{80} = \frac{\cancel{80}^{1}x}{\cancel{80}_{1}}$$
$$4.375 \text{ feet} = x$$

Julie needs to sit 4.375 feet from the fulcrum.

Evaluation: Check the solution:

$$W_1L_1 + W_2L_2 = w_1l_1 + w_2l_2$$
$$95(7.5) + 85(5) = 80(4.375) + 105(7.5)$$
$$712.5 + 425 = 350 + 787.5$$
$$1,137.5 = 1,137.5$$

TRY THESE

1. Matt weighs 110 pounds and sits four feet from the fulcrum of a seesaw. If Jean weighs 80 pounds, how far should she sit from the fulcrum to balance the seesaw?

2. At one end of a lever is a 15-pound weight which is 10 inches from the fulcrum. How much weight should be placed on the other end 12 inches from the fulcrum to balance the lever?

3. A person places a lever under a 100-pound rock that is 2.5 feet from the fulcrum. How much pressure in pounds must the person place on the other end of the lever if it is 4 feet from the fulcrum to lift the rock?

4. Where should the fulcrum be placed under a eight-foot lever if there is a 32-pound weight on one end and a 40-pound weight on the other end in order to balance the lever?

5. On a 12-foot seesaw, Kelly, weighing 96 pounds, sits on one end. Peggy, weighing 84 pounds, sits in front of her, four feet from the fulcrum which is in the center of the seesaw. On the other side at the end sits Fran, who weighs 72 pounds. Where should Carol, who weighs 100 pounds, sit in order to balance the seesaw?

SOLUTIONS

1. Let x = the distance Jean should sit from the fulcrum.

$$WL = wl$$
$$110(4) = 80(x)$$
$$440 = 80x$$
$$\frac{440}{80} = \frac{\cancel{80}^{1}x}{\cancel{80}_{1}}$$
$$5.5 = x$$

Jean should sit 5.5 feet from the fulcrum.

2. Let x = the weight placed on the other side of the lever.

$$WL = wl$$
$$15(10) = x(12)$$
$$150 = 12x$$
$$\frac{150}{12} = \frac{\cancel{12}^{1}x}{\cancel{12}_{1}}$$
$$12.5 = x$$

A weight of 12.5 pounds should be placed 12 inches from the fulcrum to balance the lever.

3. Let x = the pressure in pounds needed to lift the 100-pound rock.

$$WL = wl$$
$$100(2.5) = x(4)$$
$$250 = 4x$$
$$\frac{250}{4} = \frac{\cancel{4}^{1}x}{\cancel{4}_{1}}$$
$$62.5 = x$$

It will take 62.5 pounds of pressure to lift the rock.

4. Let x = the distance from the fulcrum where a 32-pound weight sits and $8 - x$ = the distance from the fulcrum the 40-pound weight sits.

$$WL = wl$$
$$32(x) = 40(8 - x)$$
$$32x = 320 - 40x$$
$$32x + 40x = 320 - 40x + 40x$$

$$72x = 320$$

$$\frac{\cancel{72}^{\,1}x}{\cancel{72}^{\,1}} = \frac{320}{72}$$

$$x = 4.44 \text{ feet (rounded)}$$

The fulcrum should be placed 4.44 feet from the end of the lever that has the 32-pound weight.

5. **Let x = the distance from the fulcrum where Carol should sit.**

$$W_1L_1 + W_2L_2 = w_1l_1 + w_2l_2$$

$$96(6) + 84(4) = 72(6) + (100)x$$

$$576 + 336 = 432 + 100x$$

$$912 = 432 + 100x$$

$$912 - 432 = 432 - 432 + 100x$$

$$480 = 100x$$

$$\frac{480}{100} = \frac{\cancel{100}^{\,1}x}{\cancel{100}^{\,1}}$$

$$4.8 \text{ feet} = x$$

Carol should sit 4.8 feet from the fulcrum.

In this section, you learned how to solve word problems involving levers. The basic formula is $WL = wl$. In other words, if the lever is to be balanced, the weights must be properly placed at specific lengths from the fulcrum in order to accomplish this.

Work Problems

Work problems involve people doing a job. For example, if Frank can cut a lawn in two hours and his younger brother can cut the same lawn in three hours, how long will it take them to cut the grass if they both work together? In this case, we have two people doing the same job at the same time but at different rates.

Another type of problem involves pipes filling or draining bodies of water such as tanks, reservoirs, or swimming pools at different rates. For example, if one pipe can fill a large tank in five hours, and a smaller pipe can fill the tank in three hours, how long would it take to fill the tank if both pipes are turned on at the same time? Again, we have two pipes doing the same job at different rates.

The basic principle is that the amount of work done by one person, machine, or pipe plus the amount of work done by the second person, machine, or pipe is equal to the total amount of work done in a given specific time. Also the amount of work done by a single person, machine, or pipe is equal to the rate times the time. That is,

$$\text{Rate} \times \text{Time} = \text{Amount of work done}$$

 EXAMPLE

Pete can complete a job in four hours, and Matt can do the same job in six hours. How long will it take them if they both work together at the job?

SOLUTION

Goal: You are being asked to find the time in hours it will take both people to complete the job if they work together.

Strategy: Let x = the time it takes them if they work together. Now, in one hour, Pete can complete $\frac{1}{4}$ of the job and Matt can complete $\frac{1}{6}$ of the job.

	Rate	×	Time	=	Amount of work done
Pete	$\frac{1}{4}$		x		$\frac{1}{4}x$
Matt	$\frac{1}{6}$		x		$\frac{1}{6}x$

Pete does $\frac{1}{4}x$ or $\frac{x}{4}$ amount of work and Matt does $\frac{1}{6}x$ or $\frac{x}{6}$ amount of work. These are the fractional parts of work done by each. Then the total amount of work done is 100% or 1. The equation is

$$\frac{x}{4} + \frac{x}{6} = 1$$

Implementation: Solve the equation:

The LCD of 4 and 6 is 12, so clear fractions:

$$\frac{x}{4} + \frac{x}{6} = 1$$

$$\frac{\cancel{12}^{3}}{1} \cdot \frac{x}{\cancel{4}} + \frac{\cancel{12}^{2}}{1} \cdot \frac{x}{\cancel{6}} = 12 \cdot 1$$

$$3x + 2x = 12$$

$$5x = 12$$

$$\frac{\cancel{5}^{1} x}{\cancel{5}^{1}} = \frac{12}{5}$$

$$x = 2.4 \text{ hours}$$

Hence, if both work together, they can complete the job in 2.4 hours.

Evaluation: Check the solution:

$$\frac{x}{4} + \frac{x}{6} = 1$$

$$\frac{2.4}{4} + \frac{2.4}{6} = 1$$

$$0.6 + 0.4 = 1$$

$$1 = 1$$

EXAMPLE

One pipe can fill a large tank in 10 hours and another pipe can fill a tank in 6 hours. How long will it take both pipes to fill the tank if they are turned on at the same time?

SOLUTION

Goal: You are being asked to find the time in hours it would take to fill the tank if both pipes are filling the tank at the same time.

Strategy: Let $x =$ the time it takes to fill the tank with both pipes. In one hour, the first pipe does $\frac{1}{10}$ of the work and the second pipe does $\frac{1}{6}$ of the work.

	Rate	×	Time	=	Amount of work done
First pipe	$\dfrac{1}{10}$		x		$\dfrac{1}{10}x$
Second pipe	$\dfrac{1}{6}$		x		$\dfrac{1}{6}x$

Again, the total amount of work done is 100% or 1.

The equation is

$$\frac{1}{10}x + \frac{1}{6}x = 1$$

Implementation: **Solve the equation:**

$$\frac{x}{10} + \frac{x}{6} = 1$$

The LCD is 30.

$$\frac{\cancel{30}^{3}}{1} \cdot \frac{x}{\cancel{10}^{1}} + \frac{\cancel{30}^{5}}{1} \cdot \frac{x}{\cancel{6}^{1}} = 30 \cdot 1$$

$$3x + 5x = 30$$

$$8x = 30$$

$$\frac{\cancel{8}^{1}x}{\cancel{8}^{1}} = \frac{30}{8}$$

$$x = 3.75 \text{ hours}$$

Hence, if both pipes are turned on at the same time, it would take 3.75 hours.

Evaluation: **Check the solution:**

$$\frac{x}{10} + \frac{x}{6} = 1$$

$$\frac{3.75}{10} + \frac{3.75}{6} = 1$$

$$0.375 + 0.625 = 1$$

$$1 = 1$$

As you can see, both types of problems can be done using the same strategy. The next examples show some variations of work problems.

EXAMPLE

A person can paint a meeting room in 8 hours and her assistant can paint the same room in 12 hours. If on a certain day, the assistant shows up two hours late and starts to work, how long will it take both people to paint the room?

SOLUTION

Goal: You are being asked to find the time it takes both workers to paint the room.

Strategy: Let x = the time it takes to paint the rest of the room when both people are working.

	Rate	×	Time	=	Amount of work done
First painter	$\dfrac{1}{8}$		x		$\dfrac{1}{8}x$
Assistant painter	$\dfrac{1}{12}$		x		$\dfrac{1}{12}x$

Since the assistant starts two hours later, the first painter has already done $2 \cdot \dfrac{1}{8}$ or $\dfrac{2}{8}$ of the work; hence, the equation is

$$\frac{2}{8} + \frac{1}{8}x + \frac{1}{12}x = 1$$

Implementation: Solve the equation:

$$\frac{2}{8} + \frac{1}{8}x + \frac{1}{12}x = 1$$

The LCD is 24.

$$\frac{\cancel{24}^3}{1} \cdot \frac{2}{\cancel{8}^1} + \frac{\cancel{24}^3}{1} \cdot \frac{x}{\cancel{8}^1} + \frac{\cancel{24}^2}{1} \cdot \frac{x}{\cancel{12}^1} = 24 \cdot 1$$

$$6 + 3x + 2x = 24$$

$$6 + 5x = 24$$

$$6 - 6 + 5x = 24 - 6$$

$$\frac{\cancel{5}^1 x}{\cancel{5}^1} = \frac{18}{5}$$

$$x = 3.6 \text{ hours}$$

Since the first painter has already worked two hours, the time it takes to paint the whole room is $2 + 3.6 = 5.6$ hours.

Evaluation: Check the solution:

$$\frac{2}{8} + \frac{1}{8}x + \frac{1}{12}x = 1$$

$$\frac{2}{8} + \frac{3.6}{8} + \frac{3.6}{12} = 1$$

$$0.25 + 0.45 + 0.3 = 1$$

$$1 = 1$$

EXAMPLE _____

A large water tank can be filled in 12 hours and drained in 30 hours. How long will it take to fill the tank if the owner has forgotten to close the drain valve?

SOLUTION _____

Goal: You are being asked how long in hours it will take to fill the tank if the drain is left open.

Strategy: Let x = the time in hours it takes to fill the tank.

	Rate ×	Time =	Amount of work done
Fill tank	$\frac{1}{12}$	x	$\frac{1}{12}x$
Empty tank	$\frac{1}{30}$	x	$\frac{1}{30}x$

Since the drain is emptying the tank, the equation is

$$\frac{1}{12}x - \frac{1}{30}x = 1$$

Implementation: Solve the equation:

$$\frac{1}{12}x - \frac{1}{30}x = 1$$

The LCD is 60.

$$\frac{\cancel{60}^{5}}{1} \cdot \frac{x}{\cancel{12}^{1}} - \frac{\cancel{60}^{2}}{1} \cdot \frac{x}{\cancel{30}^{1}} = 60 \cdot 1$$

$$5x - 2x = 60$$

$$3x = 60$$

$$\frac{\cancel{3}^{1}x}{\cancel{3}^{1}} = \frac{60}{3}$$

$$x = 20 \text{ hours}$$

Hence, it will take 20 hours to fill the tank.

Evaluation: Check the solution:

$$\frac{x}{12} - \frac{x}{30} = 1$$

$$\frac{20}{12} - \frac{20}{30} = 1$$

$$1.67 - 0.67 = 1$$

$$1 = 1$$

EXAMPLE

Sarah can do a job in 40 minutes and, working with Millie, both can do the job in 15 minutes. How long will it take Millie to do the job alone?

SOLUTION

Goal: You are being asked to find the time in minutes it takes for Millie to complete the job alone.

Strategy: Let x = the time it takes Millie to complete the job.

	Rate ×	Time =	Amount of work done
Sarah	$\frac{1}{40}$	15	$\frac{15}{40}$
Millie	$\frac{1}{x}$	15	$\frac{15}{x}$

The equation is

$$\frac{15}{40} + \frac{15}{x} = 1$$

Implementation: Solve the equation:

$$\frac{15}{40} + \frac{15}{x} = 1$$

The LCD = 40x.

$$\frac{\cancel{40}^{1}x}{1}\cdot\frac{15}{\cancel{40}}+\frac{40\cancel{x}^{1}}{1}\cdot\frac{15}{\cancel{x}^{1}}=40x\cdot1$$

$$x\cdot15+40\cdot15=40x$$

$$15x+600=40x$$

$$15x-15x+600=40x-15x$$

$$600=25x$$

$$\frac{600}{25}=\frac{\cancel{25}^{1}x}{\cancel{25}^{1}}$$

$$24=x$$

Hence, it will take Millie 24 minutes to do the job alone.

Evaluation: Check the solution:

$$\frac{15}{40}+\frac{15}{x}=1$$

$$\frac{15}{40}+\frac{15}{24}=1$$

$$0.375+0.625=1$$

$$1=1$$

TRY THESE

1. One pipe can empty a pool in 90 minutes, while a second pipe can empty it in 120 minutes. If both pipes are opened at the same time, how long will it take to drain the pool?

2. Joe can complete a project in 45 minutes and his brother Clem can complete it in 60 minutes. If they both work on the project at the same time, how long will it take them to complete the project?

3. Melissa can clean a barn in 4.5 hours and her father can clean it in 3 hours. How long will it take if they both work together?

4. Sam can plow a field in 6 hours and his brother Bill can plow it in 7.5 hours. How long will it take them to plow it if they use two plows and work together?

5. Sid can complete a job in 150 minutes, and if Sid and Bret both work on the job, they can complete it in 90 minutes. How long will it take Bret to complete the job by himself?

6. Pipe A can fill a tank in 12 minutes. Pipe B can fill it in 16 minutes, and pipe C can fill it in 18 minutes. If all three pipes are opened at the same time, how long will it take to fill the tank?

7. A pipe can fill a tank in 60 minutes, while the drain can drain it in 75 minutes. If the drain is left open and the fill pipe is turned on, how long will it take to fill the tank?

8. One faucet can fill a large tub in 64 minutes, while another faucet can fill the tub in 96 minutes. How long will it take to fill the tub if both faucets are opened at the same time?

9. Carl can seed a large field in four hours. His son can do the job in three hours. If the son starts an hour after his father, how long will it take to seed the field?

10. Carol can make a costume twice as fast as Ben can. If they both work together, they can make it in three hours. How long will it take Carol to make the costume if she works alone?

✔ SOLUTIONS

1. Let x = the time it takes to empty the pool if both pipes are open.

$$\frac{1}{120}x + \frac{1}{90}x = 1 \qquad LCD = 360$$

$$\frac{\overset{3}{\cancel{360}}}{1} \cdot \frac{1}{\underset{1}{\cancel{120}}}x + \frac{\overset{4}{\cancel{360}}}{1} \cdot \frac{1}{\underset{1}{\cancel{90}}}x = 360 \cdot 1$$

$$3x + 4x = 360$$

$$7x = 360$$

$$\frac{\overset{1}{\cancel{7}}x}{\underset{1}{\cancel{7}}} = \frac{360}{7}$$

$$x = 51\frac{3}{7} \text{ or } 51.43 \text{ minutes (rounded)}$$

2. Let x = the time it takes both to complete the project.

$$\frac{1}{60}x + \frac{1}{45}x = 1 \qquad \text{LCD} = 360$$

$$\frac{\overset{6}{\cancel{360}}}{1} \cdot \frac{1}{\cancel{60}^{1}}x + \frac{\overset{8}{\cancel{360}}}{1} \cdot \frac{1}{\cancel{45}^{1}}x = 360 \cdot 1$$

$$6x + 8x = 360$$

$$14x = 360$$

$$\frac{\overset{1}{\cancel{14}}x}{\cancel{14}^{1}} = \frac{360}{14}$$

$$x = 25\frac{5}{7} \text{ or } 25.71 \text{ minutes (rounded)}$$

3. Let x = the time it takes both people to complete the project if they work on it together.

$$\frac{1}{4.5}x + \frac{1}{3}x = 1 \qquad \text{LCD} = 9$$

$$\frac{\overset{2}{\cancel{9}}}{1} \cdot \frac{1}{\cancel{4.5}^{1}}x + \frac{\overset{3}{\cancel{9}}}{1} \cdot \frac{1}{\cancel{3}^{1}}x = 9 \cdot 1$$

$$2x + 3x = 9$$

$$5x = 9$$

$$\frac{\overset{1}{\cancel{5}}x}{\cancel{5}^{1}} = \frac{9}{5}$$

$$x = 1.8 \text{ hours}$$

4. Let x = the time it will take Sam and Bill to plow the field if they both work on it together.

$$\frac{1}{6}x + \frac{1}{7.5}x = 1 \qquad \text{LCD} = 30$$

$$\frac{\overset{5}{\cancel{30}}}{1} \cdot \frac{1}{\cancel{6}^{1}}x + \frac{\overset{4}{\cancel{30}}}{1} \cdot \frac{1}{\cancel{7.5}^{1}}x = 30 \cdot 1$$

$$5x + 4x = 30$$

$$9x = 30$$

$$\frac{\overset{1}{\cancel{9}}x}{\cancel{9}^{1}} = \frac{30}{9}$$

$$x = 3\frac{1}{3} \text{ or } 3.3 \text{ hours (rounded)}$$

5. Let x = the time it takes Bret to complete the job.

$$\frac{90}{150} + \frac{90}{x} = 1 \qquad \text{LCD} = 150x$$

$$\frac{\cancel{150}^{1} x}{1} \cdot \frac{90}{\cancel{150}_{1}} + \frac{150 \cancel{x}^{1}}{1} \cdot \frac{90}{\cancel{x}^{1}} = 150x \cdot 1$$

$$90x + 13{,}500 = 150x$$

$$90x - 90x + 13{,}500 = 150x - 90x$$

$$13{,}500 = 60x$$

$$\frac{13{,}500}{60} = \frac{\cancel{60}^{1} x}{\cancel{60}^{1}}$$

$$225 \text{ minutes} = x$$

6. Let x = the time it takes all three pipes to fill the tank.

$$\frac{1}{12}x + \frac{1}{16}x + \frac{1}{18}x = 1 \qquad \text{LCD} = 144$$

$$\frac{\cancel{144}^{12}}{1} \cdot \frac{1}{\cancel{12}^{1}}x + \frac{\cancel{144}^{9}}{1} \cdot \frac{1}{\cancel{16}^{1}}x + \frac{\cancel{144}^{8}}{1} \cdot \frac{1}{\cancel{18}^{1}}x = 144 \cdot 1$$

$$12x + 9x + 8x = 144$$

$$29x = 144$$

$$\frac{\cancel{29}^{1} x}{\cancel{29}^{1}} = \frac{144}{29}$$

$$x = 4\frac{28}{29} \text{ or } 4.97 \text{ minutes} \atop \text{(rounded)}$$

7. Let x = the time it takes to fill the tank.

$$\frac{1}{60}x - \frac{1}{75}x = 1 \qquad \text{LCD} = 300$$

$$\frac{\cancel{300}^{5}}{1} \cdot \frac{1}{\cancel{60}^{1}}x - \frac{\cancel{300}^{4}}{1} \cdot \frac{1}{\cancel{75}^{1}}x = 300 \cdot 1$$

$$5x - 4x = 300$$

$$x = 300 \text{ minutes}$$

8. Let x = the time it takes to fill the tub if both faucets are on.

$$\frac{1}{64}x + \frac{1}{96}x = 1 \qquad LCD = 192$$

$$\frac{\cancel{192}^3}{1} \cdot \frac{1}{\cancel{64}^1}x + \frac{\cancel{192}^2}{1} \cdot \frac{1}{\cancel{96}^1}x = 192 \cdot 1$$

$$3x + 2x = 192$$

$$5x = 192$$

$$\frac{\cancel{5}^1 x}{\cancel{5}^1} = \frac{192}{5}$$

$$x = 38.4 \text{ minutes}$$

9. Let x = the time it takes both people to seed the field. Since his father already had worked one hour before his son started, he did $\frac{1}{4}$ of the work.

$$\frac{1}{4} + \frac{1}{4}x + \frac{1}{3}x = 1 \qquad LCD = 12$$

$$\frac{\cancel{12}^3}{1} \cdot \frac{1}{\cancel{4}^1} + \frac{\cancel{12}^3}{1} \cdot \frac{1}{\cancel{4}^1}x + \frac{\cancel{12}^4}{1} \cdot \frac{1}{\cancel{3}^1}x = 12 \cdot 1$$

$$3 + 3x + 4x = 12$$

$$3 + 7x = 12$$

$$3 - 3 + 7x = 12 - 3$$

$$7x = 9$$

$$\frac{\cancel{7}^1 x}{\cancel{7}^1} = \frac{9}{7}$$

$$x = 1\frac{2}{7} \text{ or } 1.29 \text{ hours (rounded)}$$

10. Let x = the time it takes Carol to make the costume and $2x$ = the time it takes Ben to make the costume.

$$\frac{3}{x} + \frac{3}{2x} = 1 \qquad LCD = 2x$$

$$\frac{\cancel{2x}^1}{1} \cdot \frac{3}{\cancel{x}^1} + \frac{\cancel{2x}^1}{1} \cdot \frac{3}{\cancel{2x}^1} = 2x \cdot 1$$

$$6 + 3 = 2x$$

$$9 = 2x$$

$$\frac{9}{2} = \frac{\cancel{2}^{1}x}{\cancel{2}^{1}}$$

$$4.5 \text{ hours} = x$$

In this section, you have learned how to solve problems related to some kind of work. The basic formula is Rate × Time = Amount of work done.

Summary

This chapter explained how to solve finance, lever, and work problems.

QUIZ

1. A person has $15,000 invested at 6% and another sum invested at 4%. If the total interest he received on both investments was $1,700, find the amount of money he has invested at 4%.

   A. $18,000
   B. $20,000
   C. $17,000
   D. $14,000

2. An investor has some money invested at 7% and some money invested at 3%. The total interest on both investments is $518. If the total amount of money he has invested is $9,000, find the amount he has invested at 7%.

   A. $6,000
   B. $6,200
   C. $2,800
   D. $3,000

3. An investor invested $40,000, some at 5% and some at 9%. The annual interest on the 9% investment is $2,480 more than the interest on the 5% investment. How much money was invested at 9%?

   A. $24,000
   B. $28,000
   C. $30,000
   D. $32,000

4. A person has three times the amount of money invested at 4% than she has invested at 2%. If the total interest is $420, how much money is invested at 2%?

   A. $2,000
   B. $5,000
   C. $3,000
   D. $9,000

5. An 85-pound weight is placed on a board 2 feet from the fulcrum. How far from the fulcrum must an 80-pound weight be placed in order to balance the seesaw?

   A. 3.325 feet
   B. 2.625 feet
   C. 2.875 feet
   D. 2.125 feet

6. A 120-pound weight is placed on an 8-foot board with the fulcrum at the center. How much weight should be placed 3 feet from the fulcrum to balance the lever?

   A. 160 pounds
   B. 155 pounds
   C. 170 pounds
   D. 140 pounds

7. If a 120-pound weight is placed at the end of a 12-foot lever and a 150-pound weight is placed on the other end, how many feet from the 120-pound weight should the fulcrum be placed in order to balance the lever?

A. $5\dfrac{3}{4}$ feet

B. $7\dfrac{1}{3}$ feet

C. $6\dfrac{2}{3}$ feet

D. $6\dfrac{1}{4}$ feet

8. Mary can detail an automobile in 3 hours. If she gets help from her sister, they can detail the car in 1.8 hours. How long will it take her sister to detail the automobile if she works by herself?

A. 4 hours
B. 4.5 hours
C. 3.5 hours
D. 3 hours

9. Joyce can clean the windows of a building in 8 hours. Her partner can clean the same windows in 4.8 hours. How long will it take them to clean the windows of the building if they both work together?

A. 5.4 hours
B. 6 hours
C. 3 hours
D. 4.2 hours

10. A small pipe can drain a tank in 40 minutes and a large pipe can drain it in 24 minutes. If both pipes are opened at the same time, how long will it take to drain the tank?

A. 20 minutes
B. 10 minutes
C. 12 minutes
D. 15 minutes

chapter **10**

Solving Word Problems Using Two Equations

Many word problems in algebra can be solved by using two equations with two unknowns (usually x and y). When you use two unknowns, you let x = one of the unknowns and y = the other unkown. Then you can write two equations and solve them as a system of equations. Each problem will have two solutions, one for the value of x and one for the value of y.

CHAPTER OBJECTIVES

In this chapter, you will learn how to

- Solve a system of two equations
- Solve word problems using two equations

Refresher V: Systems of Equations

Two equations with two variables, usually x and y, are called a *system* of equations. For example,

$$x - y = 3$$
$$2x + y = 12$$

is called a system of equations. The solution to a system of equations consists of the values for the two variables which, when substituted in the equations, make both equations true at the same time. In this case, the solution for the system shown is $x = 5$ and $y = 2$. This can be shown as follows:

$$x - y = 3 \qquad\qquad 2x + y = 12$$
$$5 - 2 = 3 \qquad\qquad 2(5) + 2 = 12$$
$$3 = 3 \qquad\qquad 12 = 12$$

In other words, in order to solve a system of equations, it is necessary to find a value for x and a value for y which, when substituted in the equations, makes them both true.

There are several ways to solve a system of equations. The method used here is called the *substitution* method. You can use these steps:

Step 1 Select one equation and solve it for one variable in terms of the other variable.

Step 2 Substitute this expression for the variable in the other equation and solve it for the remaining variable.

Step 3 Select one of the equations, substitute the value for the variable found in Step 2, and solve for the other variable.

EXAMPLE

Solve the system:

$$3x - y = 5$$
$$x + 2y = 18$$

SOLUTION

Step 1: Select the second equation and solve it for *x* in terms of *y*.

$$x + 2y = 18$$
$$x + 2y - 2y = 18 - 2y$$
$$x = 18 - 2y$$

Step 2: Substitute 18 – 2*y* for *x* in the first equation and solve for *y*.

$$3x - y = 5$$
$$3(18 - 2y) - y = 5$$
$$54 - 6y - y = 5$$
$$54 - 7y = 5$$
$$54 - 54 - 7y = 5 - 54$$
$$-7y = -49$$
$$\frac{\cancel{-7}^{\,1}\,y}{\cancel{-7}_{\,1}} = \frac{-49}{-7}$$
$$y = 7$$

Step 3: Select 3*x* – *y* = 5, substitute *y* = 7 and solve for *x*.

$$3x - y = 5$$
$$3x - 7 = 5$$
$$3x - 7 + 7 = 5 + 7$$
$$3x = 12$$
$$\frac{\cancel{3}^{\,1}\,x}{\cancel{3}_{\,1}} = \frac{12}{3}$$
$$x = 4$$

Hence, the solution to the system is $x = 4$ and $y = 7$.

You can check the solution by substituting $x = 4$ and $y = 7$ in the other equation and see if it is true.

$$x + 2y = 18$$
$$4 + 2(7) = 18$$
$$4 + 14 = 18$$
$$18 = 18$$

EXAMPLE

Solve the system:

$$x + 4y = 3$$
$$4x - 3y = -26$$

SOLUTION

Step 1: Solve the first equation for x.

$$x + 4y = 3$$
$$x + 4y - 4y = 3 - 4y$$
$$x = 3 - 4y$$

Step 2: Substitute $3 - 4y$ for x in the second equation and solve for y.

$$4x - 3y = -26$$
$$4(3 - 4y) - 3y = -26$$
$$12 - 16y - 3y = -26$$
$$12 - 19y = -26$$
$$12 - 12 - 19y = -26 - 12$$
$$-19y = -38$$
$$\frac{\cancel{-19}^{1}y}{\cancel{-19}^{1}} = \frac{-38}{-19}$$
$$y = 2$$

Step 3: Substitute 2 for y in the first equation and find the value for x.

$$x + 4y = 3$$
$$x + 4(2) = 3$$
$$x + 8 = 3$$
$$x + 8 - 8 = 3 - 8$$
$$x = -5$$

You can check the solution by using $4x - 3y = -26$ when $x = -5$ and $y = 2$.

$$4x - 3y = -26$$
$$4(-5) - 3(2) = -26$$
$$-20 - 6 = -26$$
$$-26 = -26$$

Still Struggling

When selecting an equation and a variable to solve for in Step 1, you should look for an equation that has a variable whose numerical coefficient is 1. Since this is not always possible, you can still use the substitution method to solve the equation as shown in the next example.

EXAMPLE

Solve the system:

$$3x - 5y = -7$$
$$2x + 3y = -11$$

SOLUTION

Step 1: Select the second equation and solve for y.

$$2x + 3y = -11$$
$$2x - 2x + 3y = -11 - 2x$$
$$3y = -11 - 2x$$
$$\frac{\cancel{3}^1 y}{\cancel{3}^1} = \frac{-11 - 2x}{3}$$
$$y = \frac{-11 - 2x}{3}$$

Step 2: Substitute in the first equation.

$$3x - 5y = -7$$

$$3x - 5 \cdot \frac{-11 - 2x}{3} = -7$$

Clear fractions.

$$3 \cdot 3x - \frac{\cancel{3}^{1}}{1} \cdot \frac{5(-11 - 2x)}{\cancel{3}^{1}} = 3(-7)$$

$$9x - 5(-11 - 2x) = -21$$

$$9x + 55 + 10x = -21$$

$$19x + 55 = -21$$

$$19x + 55 - 55 = -21 - 55$$

$$19x = -76$$

$$\frac{\cancel{19}^{1} x}{\cancel{19}^{1}} = \frac{-76}{19}$$

$$x = -4$$

Step 3: Find y.

$$3x - 5y = -7$$

$$3(-4) - 5y = -7$$

$$-12 - 5y = -7$$

$$-12 + 12 - 5y = -7 + 12$$

$$-5y = 5$$

$$\frac{\cancel{-5}^{1} y}{\cancel{-5}^{1}} = \frac{5}{-5}$$

$$y = -1$$

The solution is $x = -4$ and $y = -1$. You can check the answer.

? Still Struggling

It doesn't matter which equation you use or which variable you solve for first.

TRY THESE

Solve each system.

1. $2x - y = -15$

   $x + 3y = 3$

2. $3x - 2y = -1$

   $x + y = 13$

3. $x - y = 6$

   $5x + y = 6$

4. $8x = y$

   $2x + y = 10$

5. $3x - 2y = -13$

   $-2x + 5y = -17$

SOLUTIONS

1. $2x - y = -15$

   $x + 3y = 3$

   Solve the second equation for x.

   $$x + 3y = 3$$
   $$x + 3y - 3y = 3 - 3y$$
   $$x = 3 - 3y$$

   Substitute in the first equation and solve for y.

   $$2x - y = -15$$
   $$2(3 - 3y) - y = -15$$
   $$6 - 6y - y = -15$$
   $$6 - 7y = -15$$
   $$6 - 6 - 7y = -15 - 6$$
   $$-7y = -21$$
   $$\frac{\cancel{-7}^{1}y}{\cancel{-7}^{1}} = \frac{-21}{-7}$$
   $$y = 3$$

Find x.

$$2x - y = -15$$
$$2x - 3 = -15$$
$$2x - 3 + 3 = -15 + 3$$
$$2x = -12$$
$$\frac{\cancel{2}^{1}x}{\cancel{2}^{1}} = \frac{-12}{2}$$
$$x = -6$$

The solution is $x = -6$ and $y = 3$.

2. $3x - 2y = -1$

   $x + y = 13$

Solve the second equation for x.

$$x + y = 13$$
$$x + y - y = 13 - y$$
$$x = 13 - y$$

Substitute in the first equation and solve for y.

$$3x - 2y = -1$$
$$3(13 - y) - 2y = -1$$
$$39 - 3y - 2y = -1$$
$$39 - 5y = -1$$
$$39 - 39 - 5y = -1 - 39$$
$$-5y = -40$$
$$\frac{\cancel{-5}^{1}y}{\cancel{-5}^{1}} = \frac{-40}{-5}$$
$$y = 8$$

Find x.

$$x + y = 13$$
$$x + 8 = 13$$
$$x + 8 - 8 = 13 - 8$$
$$x = 5$$

3. $x - y = 6$

$5x + y = 6$

Solve the second equation for y.

$$5x + y = 6$$

$$5x - 5x + y = 6 - 5x$$

$$y = 6 - 5x$$

Substitute in the first equation and solve for x.

$$x - y = 6$$

$$x - (6 - 5x) = 6$$

$$x - 6 + 5x = 6$$

$$6x - 6 = 6$$

$$6x - 6 + 6 = 6 + 6$$

$$6x = 12$$

$$\frac{\cancel{6}^1 x}{\cancel{6}^1} = \frac{12}{6}$$

$$x = 2$$

Find y.

$$x - y = 6$$

$$2 - y = 6$$

$$2 - 2 - y = 6 - 2$$

$$-y = 4$$

$$\frac{\cancel{-1}^1 y}{\cancel{-1}^1} = \frac{4}{-1}$$

$$y = -4$$

The solution is $x = 2$ and $y = -4$.

4. $8x = y$

$2x + y = 10$

Substitute for y in the second equation and find x since $8x = y$.

$$2x + y = 10$$

$$2x + 8x = 10$$

$$10x = 10$$

$$\frac{\cancel{10}^{1}x}{\cancel{10}^{1}} = \frac{10}{10}$$

$$x = 1$$

Find y.

$$8x = y$$

$$8(1) = y$$

$$8 = y$$

The solution is $x = 1$ and $y = 8$.

5. $3x - 2y = -13$

$-2x + 5y = -17$

Solve for x in the first equation.

$$3x - 2y = -13$$

$$3x - 2y + 2y = -13 + 2y$$

$$3x = -13 + 2y$$

$$\frac{\cancel{3}^{1}x}{\cancel{3}^{1}} = \frac{-13 + 2y}{3}$$

$$x = \frac{-13 + 2y}{3}$$

Substitute for x in the second equation.

$$-2x + 5y = -17$$

$$-2\left(\frac{-13+2y}{3}\right) + 5y = -17$$

$$\frac{26-4y}{3} + 5y = -17$$

$$\frac{\cancel{3}^{1}}{1} \cdot \frac{26-4y}{\cancel{3}^{1}} + 3 \cdot 5y = 3(-17)$$

$$26 - 4y + 15y = -51$$

$$26 + 11y = -51$$

$$26 - 26 + 11y = -51 - 26$$

$$11y = -77$$

$$\frac{\cancel{11}^{1}\,y}{\cancel{11}^{1}} = \frac{-77}{11}$$

$$y = -7$$

Find x.

$$3x - 2y = -13$$

$$3x - 2(-7) = -13$$

$$3x + 14 = -13$$

$$3x + 14 - 14 = -13 - 14$$

$$3x = -27$$

$$\frac{\cancel{3}^{1}\,x}{\cancel{3}^{1}} = \frac{-27}{3}$$

$$x = -9$$

The solution is $x = -9$ and $y = -7$.

In this refresher, you learned how to solve a system of two equations with two unknowns. The method of solution is called substitution. There are other methods that can be used to solve these systems. You can find these methods in algebra books.

Solving Word Problems Using Two Equations

NOTE *If you need to review systems of equations, complete Refresher V.*

Many of the previous types of problems can be solved using a system of two equations with two unknowns. The strategy used to solve problems using two equations is:

Step 1 Represent one of the unknowns as x and the other unknown as y.

Step 2 Translate the information about the variables into two equations using the two unknowns.

Step 3 Solve the system of equations for x and y.

In this section, a sample of each type of problem is solved by using a system of two equations with two unknowns. You will find these problems are similar to the ones in the previous sections. This was done so that you can compare the two methods (i.e., solving a problem using one equation versus solving a problem using two equations). For some types of problems, such as lever and work problems, it is better to use one equation.

EXAMPLE

One number is 16 more than another number and the sum of the two numbers is 28. Find the numbers.

SOLUTION

Goal: You are being asked to find two numbers.

Strategy: Let x = the smaller number and y = the larger number.

Since one number is 16 more than the other number, the first equation is

$$y = x + 16$$

Since the sum of the two numbers is 28, the second equation is

$$x + y = 28$$

Implementation: Solve the system:

$$y = x + 16$$
$$x + y = 28$$

Substitute the value for y in the second equation and solve for x since $y = x + 16$.

$$x + y = 28$$
$$x + x + 16 = 28$$
$$2x + 16 = 28$$
$$2x + 16 - 16 = 28 - 16$$

$$2x = 12$$

$$\frac{\cancel{2}^{1}x}{\cancel{2}^{1}} = \frac{12}{2}$$

$$x = 6$$

Find the other number.

$$y = x + 16$$

$$y = 6 + 16$$

$$y = 22$$

Hence, the numbers are 6 and 22.

Evaluation: Check the second equation.

$$x + y = 28$$

$$6 + 22 = 28$$

$$28 = 28$$

EXAMPLE

The sum of the digits of a two-digit number is 14. If the digits are reversed, the new number is 18 more than the original number. Find the number.

 SOLUTION

Goal: You are being asked to find a two-digit number.

Strategy: Let x = the tens digit

y = the ones digit

Then

$10x + y$ = original number

$10y + x$ = new number with digits reversed

Since the sum of the digits of the number is 14, the first equation is

$$x + y = 14$$

Since reversing the digits gives a new number that is 18 more than the original number, the second equation is

$$(10x + y) + 18 = (10y + x)$$

Implementation: Solve the system:

$$x + y = 14$$
$$10x + y + 18 = 10y + x$$

Solve the first equation for y.

$$x + y = 14$$
$$x - x + y = 14 - x$$
$$y = 14 - x$$

Substitute in the second equation and find x.

$$10x + y + 18 = 10y + x$$
$$10x + 14 - x + 18 = 10(14 - x) + x$$
$$9x + 32 = 140 - 10x + x$$
$$9x + 32 = 140 - 9x$$
$$9x + 9x + 32 = 140 - 9x + 9x$$
$$18x + 32 - 32 = 140 - 32$$
$$18x = 108$$
$$\frac{\cancel{18}^{1} x}{\cancel{18}_{1}} = \frac{108}{18}$$
$$x = 6$$

Find y.

$$x + y = 14$$
$$6 + y = 14$$
$$6 - 6 + y = 14 - 6$$
$$y = 8$$

Hence, the number is 68.

Evaluation: Check the information in the second equation.

$$\text{Original number} = 68$$
$$\text{Reversed number} = 86$$

Since 86 is 18 more than 68, the answer is correct.

EXAMPLE

A person has 12 coins consisting of quarters and dimes. If the total amount of this change is $2.25, how many of each kind of coin are there?

SOLUTION

Goal: You are being asked to find how many coins are quarters and how many coins are dimes.

Strategy: Let x = the number of quarters

y = the number of dimes

$25x$ = the value of the quarters

$10y$ = the value of the dimes

Since there are 12 coins, the first equation is

$$x + y = 12$$

Since the total value of the quarters plus the dimes is $2.25 or 225¢, the second equation is

$$25x + 10y = 225$$

Implementation: Solve the system:

$$x + y = 12$$
$$25x + 10y = 225$$

Solve for y in the first equation.

$$x + y = 12$$
$$x - x + y = 12 - x$$
$$y = 12 - x$$

Substitute this expression for *y* in the second equation and solve for *x*.

$$25x + 10y = 225$$
$$25x + 10(12 - x) = 225$$
$$25x + 120 - 10x = 225$$
$$15x + 120 = 225$$
$$15x + 120 - 120 = 225 - 120$$
$$15x = 105$$
$$\frac{\cancel{15}^1 x}{\cancel{15}^1} = \frac{105}{15}$$
$$x = 7$$

Find *y*.

$$x + y = 12$$
$$7 + y = 12$$
$$7 - 7 + y = 12 - 7$$
$$y = 5$$

Hence, there are 7 quarters and 5 dimes.

Evaluation: Find the values of each and see if their sum is $2.25.

$$7 \text{ quarters} = 7 \times \$0.25 = \$1.75$$
$$5 \text{ dimes} = 5 \times \$0.10 = \$0.50$$
$$\$1.75 + \$0.50 = \$2.25$$

EXAMPLE

Sam is 10 years younger than his brother. In two years, his brother will be three times as old as Sam. Find their present ages.

SOLUTION

Goal: You are being asked to find the present ages of Sam and his brother.

Strategy: Let *x* = Sam's age

y = his brother's age

x + 2 = Sam's age in two years

y + 2 = his brother's age in two years

Since Sam is 10 years younger than his brother, the first equation is

$$x + 10 = y$$

In two years, Sam's brother will be three times as old as Sam, so the second equation is

$$y + 2 = 3(x + 2)$$

Implementation: Solve the system:

$$x + 10 = y$$
$$y + 2 = 3(x + 2)$$

Substitute the value of y in the second equation and solve for x since $x + 10 = y$.

$$y + 2 = 3(x + 2)$$
$$x + 10 + 2 = 3(x + 2)$$
$$x + 12 = 3x + 6$$
$$x - x + 12 = 3x - x + 6$$
$$12 = 2x + 6$$
$$12 - 6 = 2x + 6 - 6$$
$$6 = 2x$$
$$\frac{6}{2} = \frac{\cancel{2}^{1} x}{\cancel{2}^{1}}$$
$$3 = x$$

Select the first equation, let $x = 3$, and solve for y.

$$x + 10 = y$$
$$3 + 10 = y$$
$$13 = y$$

Hence, Sam's brother is 13 years old and Sam is 3 years old.

Evaluation: Sam's age is 3, which is 10 years younger than his brother who is 13 years old. In two years, Sam will be 5 and his brother will be 15. Hence his brother will be three times as old as Sam.

EXAMPLE

A person drove his car from home to a repair shop at 30 miles per hour and walked home at 3 miles per hour. If the total trip took 33 minutes, how far is the repair shop from his home?

SOLUTION

Goal: You are being asked to find the distance from the person's home to the repair shop.

Strategy: Let x = the time the person drove and y = the time the person walked.

Since the total time is 33 minutes or $\dfrac{33}{60} = 0.55$ hour, the first equation is

$$x + y = 0.55$$

Since the distances are equal and $D = RT$, the second equation is

$$30x = 3y$$

Implementation: Solve the system:

$$x + y = 0.55$$
$$30x = 3y$$

Solve the first equation for y and substitute the value in the second equation, and then solve for x.

$$x + y = 0.55$$
$$x - x + y = 0.55 - x$$
$$y = 0.55 - x$$

Then:

$$30x = 3y$$
$$30x = 3(0.55 - x)$$
$$30x = 1.65 - 3x$$
$$30x + 3x = 1.65 - 3x + 3x$$
$$33x = 1.65$$
$$\frac{\cancel{33}^{1} x}{\cancel{33}^{1}} = \frac{1.65}{33}$$
$$x = 0.05 \text{ hours}$$

Find the distance using $D = RT$.

$$D = RT$$
$$D = 30(0.05)$$
$$= 1.5 \text{ miles}$$

Evaluation: The time he walked is $0.55 - 0.05 = 0.5$ hours. The distance is $D = RT$.

$$D = 3(0.5)$$
$$= 1.5 \text{ miles}$$

Still Struggling

In the previous example, the rates are given in miles per hour and the total time is given in minutes, i.e., 33 minutes. Therefore, it is necessary to convert the minutes to hours so that the units in the problem are the same.

EXAMPLE

A merchant mixes some cashews costing $6 a pound with some peanuts costing $2 a pound. How much of each must be used in order to make 25 pounds of mixture costing $3.50 a pound?

SOLUTION

Goal: You are being asked to find how much of each kind of nuts should be used.

Strategy: Let $x =$ the amount of $6 cashews used and $y =$ the amount of $2 peanuts used.

Since the total amount of the mixture is 25 pounds, the first equation is

$$x + y = 25$$

Since the cost of the mixture is $3.50, the second equation is

$$6x + 2y = 25(3.50)$$

Implementation: Solve the system:

$$x + y = 25$$
$$6x + 2y = 25(3.50)$$

Solve the first equation for x. Substitute in the second equation and solve for y.

$$x + y = 25$$
$$x + y - y = 25 - y$$
$$x = 25 - y$$

Substitute:

$$6x + 2y = 25(3.50)$$
$$6(25 - y) + 2y = 25(3.50)$$
$$150 - 6y + 2y = 87.5$$
$$150 - 4y = 87.5$$
$$150 - 150 - 4y = 87.5 - 150$$
$$-4y = -62.5$$
$$\frac{\cancel{-4}^{1} y}{\cancel{-4}^{1}} = \frac{-62.5}{-4}$$
$$y = 15.625 \text{ pounds}$$

Solve for x.

$$x + y = 25$$
$$x + 15.625 = 25$$
$$x + 15.625 - 15.625 = 25 - 15.265$$
$$x = 9.375 \text{ pounds}$$

Hence, 9.375 pounds of the $6 cashews are needed and 15.625 pounds of the $2 peanuts are needed.

Evaluation: Check the second equation.

$$6x + 2y = 25(3.50)$$
$$6(9.375) + 2(15.625) = 87.5$$
$$56.25 + 31.25 = 87.5$$
$$87.5 = 87.5$$

EXAMPLE

A person has \$8,000 to invest and decides to invest part of it at 3% and the rest of it at $7\frac{1}{2}$%. If the total interest for the year is \$330, how much does the person have invested at each rate?

SOLUTION

Goal: You are being asked to find the amounts of money invested at each rate.

Strategy: Let x = the amount of money invested at 3% and y = the amount of money invested at $7\frac{1}{2}$%.

Since the total amount of money is \$8,000, the first equation is

$$x + y = \$8,000$$

Since the total interest is \$330, the second equation is

$$3\%x + 7\frac{1}{2}\%(y) = \$330$$

Implementation: Solve the system:

$$x + y = \$8,000$$
$$3\%x + 7\frac{1}{2}\%(y) = \$330$$

Solve the first equation for x. Substitute in the second equation and solve for y.

$$x + y = 8,000$$
$$x + y - y = 8,000 - y$$
$$x = 8,000 - y$$

Then:

$$0.03x + 0.075y = 330$$

$$0.03(8,000 - y) + 0.075y = 330$$

$$240 - 0.03y + 0.075y = 330$$

$$240 + 0.045y = 330$$

$$240 - 240 + 0.045y = 330 - 240$$

$$0.045y = 90$$

$$\frac{\cancel{0.045}^{1}y}{\cancel{0.045}_{1}} = \frac{90}{0.045}$$

$$y = 2,000$$

Find x.

$$x + y = 8,000$$

$$x + 2,000 = 8,000$$

$$x + 2,000 - 2,000 = 8,000 - 2,000$$

$$x = 6,000$$

Hence, the person has \$6,000 invested at 3% and \$2,000 invested at $7\frac{1}{2}\%$.

Evaluation: Check the second equation

$$3\%x + 7\frac{1}{2}\%y = \$330$$

$$3\%(\$6,000) + 7\frac{1}{2}\%(\$2,000) = \$330$$

$$0.03(6,000) + 0.075(2,000) = 330$$

$$180 + 150 = 330$$

$$330 = 330$$

TRY THESE

Use two equations with two unknowns.

1. The larger of two numbers is 12 more than the smaller number. The sum of the numbers is 50. Find the numbers.

2. An investor has $10,000 to invest at 5% and 2%. Find the amount of each investment if the total interest per year is $410.

3. Janice is twice as old as Jane, and the sum of their ages next year will be 41. Find their present ages.

4. A young person bought some apples at $1 each and sold them for $1.25 each at a flea market. His profit was $6.50. If he gave two apples to his friends, how many apples did he buy?

5. A person has 24 coins in dimes and quarters. If the total amount of money she has is $4.65, how many quarters and dimes does the person have?

6. Find two consecutive odd numbers whose sum is 88.

7. Harry bought 12 stamps. If he purchased two more 50-cent stamps than 25-cent stamps and it cost him $4.75, how many of each kind of stamps did he purchase?

8. The sum of Marci's age and her brother's age is 21. If Marci is 11 years older than her brother, find Marci's age.

9. The sum of the digits of a two-digit number is 15. If the digits are reversed, the new number is 9 less than the original number. Find the number.

10. Mr. Lee invested part of $9,500 into an account that pays 2% interest and the rest of it into an account that pays 4.5% interest. If the total interest per year he receives is $346.25, find the amount of money he has invested in each account.

SOLUTIONS

1. Let x = the larger number and y = the smaller number.

$$x = y + 12$$
$$x + y = 50$$

Substitute $y + 12$ for x in the second equation and solve for x.

$$x + y = 50$$
$$y + 12 + y = 50$$
$$2y + 12 = 50$$
$$2y + 12 - 12 = 50 - 12$$
$$2y = 38$$
$$\frac{\overset{1}{\cancel{2}}y}{\underset{1}{\cancel{2}}} = \frac{38}{2}$$
$$y = 19$$
$$x = y + 12$$
$$x = 19 + 12$$
$$x = 31$$

The larger number is 31 and the smaller number is 19.

2. **Let x = the amount of money invested at 5% and y = the amount of money invested at 2%.**

$$x + y = \$10{,}000$$
$$5\%(x) + 2\%(y) = \$410$$
$$x + y = \$10{,}000$$
$$x + y - y = \$10{,}000 - y$$
$$x = 10{,}000 - y$$
$$0.05(10{,}000 - y) + 0.02y = 410$$
$$500 - 0.05y + 0.02y = 410$$
$$500 - 0.03y = 410$$
$$500 - 500 - 0.03y = 410 - 500$$
$$-0.03y = -90$$
$$\frac{\overset{1}{\cancel{-0.03}}y}{\underset{1}{\cancel{-0.03}}} = \frac{-90}{-0.03}$$
$$y = \$3{,}000$$
$$x + y = \$10{,}000$$
$$x + 3{,}000 - 3{,}000 = 10{,}000 - 3{,}000$$
$$x = \$7{,}000$$

$7,000 should be invested at 5% and $3,000 should be invested at 2%.

3. Let x = Janice's age and y = Jane's age; then $x = 2y$ and $x + 1 + y + 1 = 41$.

$$x = 2y$$
$$x + 1 + y + 1 = 41$$
$$2y + 1 + y + 1 = 41$$
$$3y + 2 = 41$$
$$3y + 2 - 2 = 41 - 2$$
$$3y = 39$$
$$\frac{\cancel{3}^{1}y}{\cancel{3}_{1}} = \frac{39}{3}$$
$$y = 13$$
$$x = 2y$$
$$x = 2(13) = 26$$

Janice is 26 years old and Jane is 13 years old.

4. Let x = the number of apples he bought and y = the number of apples he sold.

$$x = y + 2$$
$$1.25y - 1.00x = 6.50$$
$$1.25y - 1(y + 2) = 6.50$$
$$1.25y - y - 2 = 6.50$$
$$0.25y - 2 = 6.50$$
$$0.25y - 2 + 2 = 6.50 + 2$$
$$0.25y = 8.5$$
$$\frac{\cancel{0.25}^{1}y}{\cancel{0.25}_{1}} = \frac{8.5}{0.25}$$
$$y = 34$$
$$x = y + 2 = 34 + 2 = 36$$

He bought 36 apples.

5. Let x = the number of quarters the person has and y = the number of dimes the person has.

$$x + y = 24 \text{ and } 0.25x + 0.10y = 4.65$$
$$x + y = 24$$
$$x = 24 - y$$
$$0.25(24 - y) + 0.10y = 4.65$$
$$6 - 0.25y + 0.10y = 4.65$$
$$6 - 0.15y = 4.65$$
$$6 - 6 - 0.15y = 4.65 - 6$$
$$-0.15y = -1.35$$
$$\frac{\cancel{-0.15}^{1} y}{\cancel{-0.15}^{1}} = \frac{-1.35}{-0.15}$$
$$y = 9$$
$$x + y = 24$$
$$x + 9 = 24$$
$$x + 9 - 9 = 24 - 9$$
$$x = 15$$

The person has 15 quarters and 9 dimes.

6. Let x = the first consecutive odd number and y = the second consecutive odd number.

$$y = x + 2$$
$$x + y = 88$$
$$x + x + 2 = 88$$
$$2x + 2 = 88$$
$$2x + 2 - 2 = 88 - 2$$
$$2x = 86$$
$$\frac{\cancel{2}^{1} x}{\cancel{2}^{1}} = \frac{86}{2}$$
$$x = 43$$
$$y = x + 2$$
$$y = 43 + 2$$
$$y = 45$$

The consecutive odd numbers are 43 and 45.

7. Let x = the number of 50-cent stamps and y = the number of 25-cent stamps.

$$x + y = 12$$
$$0.50x + 0.25y = 4.75$$
$$x + y = 12$$
$$x = 12 - y$$
$$0.50x + 0.25y = 4.75$$
$$0.50(12 - y) + 0.25y = 4.75$$
$$6 - 0.50y + 0.25y = 4.75$$
$$6 - 0.25y = 4.75$$
$$6 - 6 - 0.25y = 4.75 - 6$$
$$-0.25y = -1.25$$
$$\frac{\cancel{-0.25}^{\,1}\,y}{\cancel{-0.25}^{\,1}} = \frac{-1.25}{-0.25}$$
$$y = 5$$
$$x + y = 12$$
$$x + 5 = 12$$
$$x + 5 - 5 = 12 - 5$$
$$x = 7$$

Harry bought seven 50-cent stamps and five 25-cent stamps.

8. Let x = Marci's age and y = her brother's age.

$$x + y = 21$$
$$x = y + 11$$
$$x + y = 21$$
$$y + 11 + y = 21$$
$$2y + 11 = 21$$
$$2y + 11 - 11 = 21 - 11$$
$$2y = 10$$
$$y = 5$$
$$x + y = 21$$
$$x + 5 = 21$$
$$x + 5 - 5 = 21 - 5$$
$$x = 16$$

Marci is 16 years old and her brother is 5 years old.

9. Let x = the ones digit and y = the tens digit.

$$x + y = 15$$
$$10y + x - (10x + y) = 9$$
$$x + y = 15$$
$$x + y - y = 15 - y$$
$$x = 15 - y$$
$$10y + x - (10x + y) = 9$$
$$10y + 15 - y - [10(15 - y) + y] = 9$$
$$9y + 15 - [150 - 10y + y] = 9$$
$$9y + 15 - [150 - 9y] = 9$$
$$9y + 15 - 150 + 9y = 9$$
$$18y - 135 = 9$$
$$18y - 135 + 135 = 9 + 135$$
$$18y = 144$$
$$\frac{\cancel{18}^{1}\,y}{\cancel{18}^{1}} = \frac{144}{18}$$
$$y = 8$$
$$x + y = 15$$
$$x + 8 = 15$$
$$x + 8 - 8 = 15 - 8$$
$$x = 7$$

The number is 87.

10. Let x = the amount of money invested at 2% and y = the amount of money invested at 4.5%.

$$x + y = \$9,500$$
$$2\%(x) + 4.5\%(y) = \$346.25$$
$$x + y = 9,500$$
$$x + y - y = 9,500 - y$$
$$x = 9,500 - y$$
$$2\%(\$9,500 - y) + 4.5\%(y) = \$346.25$$

$$0.02(9,500 - y) + 0.045y = 346.25$$
$$190 - 0.02y + 0.045y = 346.25$$
$$190 + 0.025y = 346.25$$
$$190 - 190 + 0.025y = 346.25 - 190$$
$$0.025y = 156.25$$
$$\frac{\cancel{0.025}^{1}y}{\cancel{0.025}^{1}} = \frac{156.25}{0.025}$$
$$y = 6,250$$
$$x + y = 9,500$$
$$x + 6,250 = 9,500$$
$$x + 6,250 - 6,250 = 9,500 - 6,250$$
$$x = 3,250$$

Mr. Lee invested \$3,250 at 2% and \$6,250 at 4.5%.

Summary

In this chapter, you learned how to solve word problems using two equations with two unknowns. These equations are called a system of equations. This method is an alternative to the methods that use one equation.

QUIZ

(Use two equations to solve these problems.)

1. If the sum of two numbers is 51 and the difference is 13, find the larger number.
   A. 19
   B. 16
   C. 32
   D. 35

2. Four computers and seven printers cost $1,960, while seven computers and four printers cost $2,770. Find the cost of one computer.
   A. $180
   B. $350
   C. $600
   D. $80

3. An investor has a total of $11,000, part of which he invested at 2% interest and the rest he invested at 4.5%. If the yearly interest from the investment is $305, find the amount of money invested at 4.5%.
   A. $5,200
   B. $3,400
   C. $5,800
   D. $7,600

4. If a person can travel 10 miles upstream in 5 hours and the same distance downstream in 1.25 hours, find the rate of the current.
   A. 8 miles per hour
   B. 5 miles per hour
   C. 7 miles per hour
   D. 3 miles per hour

5. Molly has some coins in her purse. She has two more quarters than dimes and two times as many pennies as dimes. If she has a total of $1.98, how many dimes does she have?
   A. 4
   B. 5
   C. 6
   D. 8

6. A woman is five years older than her sister. Twenty years ago, she was twice as old as her sister. Find her age.
   A. 24
   B. 28
   C. 30
   D. 32

7. The sum of the digits of a two-digit number is 8. If the digits are reversed, the new number is 36 less than the original number. Find the number.

   A. 44
   B. 53
   C. 71
   D. 62

8. A grocer wants to mix some cookies costing $3 per dozen with some cookies costing $1.75 per dozen. If she wants a total of 10 dozen that sell for $2.25 per dozen, how many dozens of $3 cookies will she need?

   A. 4
   B. 3
   C. 2.25
   D. 2

9. Find the smaller of two consecutive even numbers if their sum is 86.

   A. 40
   B. 42
   C. 44
   D. 46

10. The sum of Harry's age and Larry's age is 92. Four years ago, Harry was three times as old as Larry. Find Harry's age now.

   A. 25
   B. 21
   C. 67
   D. 63

chapter 11

Solving Word Problems Using Quadratic Equations

This chapter explains how to solve word problems by using a quadratic equation or second degree equation. This equation has an x^2 term. The refresher section shows how to solve a quadratic equation by factoring.

CHAPTER OBJECTIVES

In this chapter, you will learn how to

- Solve a quadratic equation by factoring
- Solve algebra problems using quadratic equations

Refresher VI: Solving Quadratic Equations by Factoring

An equation such as $2x^2 + 3x - 5 = 0$ is called a *quadratic equation* or a *second degree equation*. There is one variable (usually x) and a second-degree term (usually x^2). There are several ways to solve quadratic equations. The method shown here will use *factoring*. *If you cannot factor trinomials, you will need to consult an algebra book to learn this skill.*

A quadratic equation can be written in *standard form* where the x^2 term is first, the x term is second, and the constant term is the third. Also, zero is on the right side of the equation. For example, the quadratic equation $2x + x^2 = 8$ can be written in standard form as $x^2 + 2x - 8 = 0$. In order to solve a quadratic equation by factoring, you should follow these steps:

Step 1 Write the equation in standard form.

Step 2 Factor the left side of the equation.

Step 3 Set both factors equal to zero.

Step 4 Solve each equation.

EXAMPLE

Find the solution to $5x + x^2 = 24$.

✔ SOLUTION

Step 1: Write the equation in standard form.

$$x^2 + 5x - 24 = 0$$

Step 2: Factor the left side.

$$(x + 8)(x - 3) = 0$$

Step 3: Set each factor to zero.

$$x + 8 = 0 \quad \text{and} \quad x - 3 = 0$$

Step 4: Solve each equation.

$$x + 8 = 0 \qquad\qquad x - 3 = 0$$

$$x + 8 - 8 = 0 - 8 \qquad x - 3 + 3 = 0 + 3$$

$$x = -8 \qquad\qquad x = 3$$

Notice that there are two solutions. You can check each value in the original equation.

$$x = -8: \quad 5x + x^2 = 24$$

$$5(-8) + (-8)^2 = 24$$

$$-40 + 64 = 24$$

$$24 = 24$$

$$x = 3: \quad 5x + x^2 = 24$$

$$5(3) + (3)^2 = 24$$

$$15 + 9 = 24$$

$$24 = 24$$

EXAMPLE

Solve $6x^2 - 24 = 7x$.

SOLUTION

Step 1: Write in standard form.

$$6x^2 - 7x - 24 = 0$$

Step 2: Factor the left side.

$$(3x - 8)(2x + 3) = 0$$

Step 3: Set both factors equal to zero.

$$3x - 8 = 0 \qquad 2x + 3 = 0$$

Step 4: Solve each equation.

$$3x - 8 = 0 \qquad\qquad 2x + 3 = 0$$

$$3x - 8 + 8 = 0 + 8 \qquad 2x + 3 - 3 = 0 - 3$$

$$3x = 8 \qquad\qquad 2x = -3$$

$$\frac{\cancel{3}^1 x}{\cancel{3}^1} = \frac{8}{3} \qquad\qquad \frac{\cancel{2}^1 x}{\cancel{2}^1} = -\frac{3}{2}$$

$$x = \frac{8}{3} \qquad\qquad x = -\frac{3}{2}$$

EXAMPLE

Solve $3x^2 = 27$.

SOLUTION

Step 1: Write in standard form.

$$3x^2 - 27 = 0$$

Step 2: Factor the left side.

$$3(x + 3)(x - 3) = 0$$

Step 3: Divide both sides by 3 and set both factors equal to zero.

$$x + 3 = 0 \qquad x - 3 = 0$$

Step 4: Solve each equation.

$$x + 3 = 0 \qquad\qquad x - 3 = 0$$
$$x + 3 - 3 = 0 - 3 \qquad x - 3 + 3 = 0 + 3$$
$$x = -3 \qquad\qquad x = 3$$

EXAMPLE

Solve $x^2 = 8x$.

SOLUTION

Step 1: $\qquad\qquad x^2 - 8x = 0$

Step 2: $\qquad\qquad x(x - 8) = 0$

Step 3: $\qquad\qquad x = 0 \qquad x - 8 = 0$

Step 4: $\qquad\qquad x = 0 \qquad x - 8 + 8 = 0 + 8$

$$x = 8$$

Quadratic equations generally have two different solutions; however, some have only one solution since the solutions are equal.

? Still Struggling

It should be noted that not all quadratic equations can be solved by factoring. However, for the purposes of this book, the solutions to the word problems in this chapter can be solved using factoring.

TRY THESE

1. $x^2 - 12 = 4x$

2. $10x = x^2 + 21$

3. $4x^2 + 19x = 5$

4. $6x^2 - 31x = -35$

5. $4x^2 = 16x$

6. $x^2 = 49$

7. $x^2 + 4x = 5$

8. $x^2 + 1 = 2x + 25$

9. $x^2 = 12x$

10. $3x^2 - 2 = 5x$

SOLUTIONS

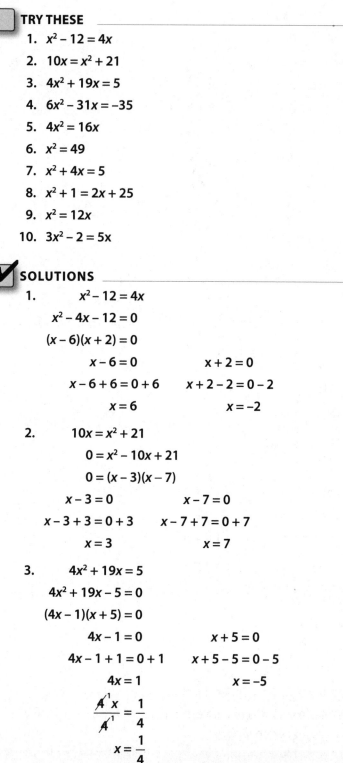

1. $x^2 - 12 = 4x$

   $x^2 - 4x - 12 = 0$

   $(x - 6)(x + 2) = 0$

   $x - 6 = 0 \qquad\qquad x + 2 = 0$

   $x - 6 + 6 = 0 + 6 \qquad x + 2 - 2 = 0 - 2$

   $x = 6 \qquad\qquad x = -2$

2. $10x = x^2 + 21$

   $0 = x^2 - 10x + 21$

   $0 = (x - 3)(x - 7)$

   $x - 3 = 0 \qquad\qquad x - 7 = 0$

   $x - 3 + 3 = 0 + 3 \qquad x - 7 + 7 = 0 + 7$

   $x = 3 \qquad\qquad x = 7$

3. $4x^2 + 19x = 5$

   $4x^2 + 19x - 5 = 0$

   $(4x - 1)(x + 5) = 0$

   $4x - 1 = 0 \qquad\qquad x + 5 = 0$

   $4x - 1 + 1 = 0 + 1 \qquad x + 5 - 5 = 0 - 5$

   $4x = 1 \qquad\qquad x = -5$

   $\dfrac{\cancel{4}^{1} x}{\cancel{4}^{1}} = \dfrac{1}{4}$

   $x = \dfrac{1}{4}$

4. $6x^2 - 31x = -35$

$6x^2 - 31x + 35 = 0$

$(3x - 5)(2x - 7) = 0$

$3x - 5 = 0 \qquad\qquad 2x - 7 = 0$

$3x - 5 + 5 = 0 + 5 \qquad 2x - 7 + 7 = 0 + 7$

$3x = 5 \qquad\qquad 2x = 7$

$\dfrac{\cancel{3}^{1}x}{\cancel{3}^{1}} = \dfrac{5}{3} \qquad\qquad \dfrac{\cancel{2}^{1}x}{\cancel{2}^{1}} = \dfrac{7}{2}$

$x = \dfrac{5}{3} \qquad\qquad x = \dfrac{7}{2}$

5. $4x^2 = 16x$

$4x^2 - 16x = 0$

$4x(x - 4) = 0$

$4x = 0 \qquad\qquad x - 4 = 0$

$\dfrac{\cancel{4}^{1}x}{\cancel{4}^{1}} = \dfrac{0}{4} \qquad x - 4 + 4 = 0$

$x = 0 \qquad\qquad x = 4$

6. $x^2 = 49$

$x^2 - 49 = 49 - 49$

$x^2 - 49 = 0$

$(x + 7)(x - 7) = 0$

$x + 7 = 0 \qquad\qquad x - 7 = 0$

$x + 7 - 7 = 0 - 7 \qquad x - 7 + 7 = 0 + 7$

$x = -7 \qquad\qquad x = 7$

7. $x^2 + 4x = 5$

$x^2 + 4x - 5 = 5 - 5$

$x^2 + 4x - 5 = 0$

$(x + 5)(x - 1) = 0$

$x + 5 = 0 \qquad\qquad x - 1 = 0$

$x + 5 - 5 = 0 - 5 \qquad x - 1 + 1 = 0 + 1$

$x = -5 \qquad\qquad x = 1$

8.
$$x^2 + 1 = 2x + 25$$
$$x^2 + 1 - 2x = 2x - 2x + 25$$
$$x^2 + 1 - 2x = 25$$
$$x^2 + 1 - 2x - 25 = 25 - 25$$
$$x^2 - 2x - 24 = 0$$
$$(x - 6)(x + 4) = 0$$

$$x - 6 = 0 \qquad\qquad x + 4 = 0$$
$$x - 6 + 6 = 0 + 6 \qquad x + 4 - 4 = 0 - 4$$
$$x = 6 \qquad\qquad x = -4$$

9.
$$x^2 = 12x$$
$$x^2 - 12x = 12x - 12x$$
$$x^2 - 12x = 0$$
$$x(x - 12) = 0$$

$$x = 0 \qquad\qquad x - 12 = 0$$
$$x - 12 + 12 = 0 + 12$$
$$x = 12$$

10.
$$3x^2 - 2 = 5x$$
$$3x^2 - 2 - 5x = 5x - 5x$$
$$3x^2 - 5x - 2 = 0$$
$$(3x + 1)(x - 2) = 0$$

$$3x + 1 = 0 \qquad\qquad x - 2 = 0$$
$$3x + 1 - 1 = 0 - 1 \qquad x - 2 + 2 = 0 + 2$$
$$3x = -1 \qquad\qquad x = 2$$
$$\frac{\cancel{3}^{1} x}{\cancel{3}^{1}} = -\frac{1}{3}$$
$$x = -\frac{1}{3}$$

In this refresher, you learned how to solve a quadratic equation by factoring. There are two other methods that are used to solve a quadratic equation. One method is completing the square. The other method is using the quadratic formula.

The quadratic formula can be used to solve all quadratic equations. It can be found in most basic algebra textbooks. When using it, you follow Steps 1 and 2 given here and then use the formula for Steps 3 and 4 to get the solution.

Solving Word Problems Using Quadratic Equations

Many problems in mathematics can be solved using a quadratic equation. The strategy you can use is:

Step 1 Represent the unknown using x and the other unknown in terms of x.

Step 2 From the problem, write expressions that are related to the unknown.

Step 3 Write the quadratic equation.

Step 4 Solve the quadratic equation for x.

Recall that a quadratic equation has two solutions. (Note: Sometimes the two solutions are equal to each other.) Both solutions can be answers to the problems; however, many times only one solution is meaningful. In that case, disregard the solution that does not make sense.

EXAMPLE

If the sum of two numbers is 18 and the product of the two numbers is 72, find the numbers.

SOLUTION

Goal: You are being asked to find two numbers whose sum is 18 and whose product is 72.

Strategy: Let x = one number and $(18 - x)$ = the other number.

If the product of the two numbers is 72, the equation is $x(18 - x) = 72$.

Implementation: Solve the equation:

$$x(18 - x) = 72$$

$$18x - x^2 = 72$$

$$0 = x^2 - 18x + 72$$

$$0 = (x - 6)(x - 12)$$

$$x - 6 = 0 \qquad\qquad x - 12 = 0$$
$$x - 6 + 6 = 0 + 6 \qquad x - 12 + 12 = 0 + 12$$
$$x = 6 \qquad\qquad x = 12$$

Hence, the two numbers are 6 and 12.

Evaluation: Check the facts of the problem. The sum 6 + 12 is 18 and the product is $6 \cdot 12 = 72$.

EXAMPLE

If the product of two consecutive numbers is 156, find the numbers.

✔ SOLUTION

Goal: You are being asked to find two consecutive numbers whose product is 156.

Strategy: Let x = the first number and $x + 1$ = the next number.

The equation for the product is $x(x + 1) = 156$.

Implementation: Solve the equation:

$$x(x + 1) = 156$$
$$x^2 + x = 156$$
$$x^2 + x - 156 = 156 - 156$$
$$(x + 13)(x - 12) = 0$$
$$x + 13 = 0 \qquad\qquad x - 12 = 0$$
$$x + 13 - 13 = 0 - 13 \qquad x - 12 + 12 = 0 + 12$$
$$x = -13 \qquad\qquad x = 12$$

Hence, the numbers are 12 and 13 or –12 and –13.

Evaluation: Find each product: $12 \cdot 13 = 156$, and $-12 \cdot (-13) = 156$

EXAMPLE

The sum of two numbers is 20. If the sum of their reciprocals is $\dfrac{5}{24}$, find the numbers.

✔ SOLUTION

Goal: You are being asked to find two numbers whose sum is 20 and whose sum of their reciprocals is $\dfrac{5}{24}$.

Strategy: Let x = one number and $20 - x$ = the other number.

The reciprocals are $\dfrac{1}{x}$ and $\dfrac{1}{20-x}$.

Then the sum of the reciprocals is $\dfrac{1}{x} + \dfrac{1}{20-x} = \dfrac{5}{24}$.

Implementation: Solve the equation:

$$\frac{1}{x} + \frac{1}{20-x} = \frac{5}{24}$$

$$LCD = 24x(20-x)$$

$$24x \cdot (20-x) \cdot \frac{1}{x} + 24x \cdot (20-x)^1 \cdot \frac{1}{(20-x)^1} = 24^1 x \cdot (20-x) \cdot \frac{5}{24^1}$$

$$24(20-x) + 24x = 5x(20-x)$$

$$480 - 24x + 24x = 100x - 5x^2$$

$$480 = 100x - 5x^2$$

$$480 + 5x^2 = 100x - 5x^2 + 5x^2$$

$$5x^2 + 480 = 100x$$

$$5x^2 - 100x + 480 = 100x - 100x$$

$$5x^2 - 100x + 480 = 0$$

Divide by 5.

$$x^2 - 20x + 96 = 0$$

$$(x - 12)(x - 8) = 0$$

$$x - 12 = 0 \qquad\qquad x - 8 = 0$$

$$x - 12 + 12 = 0 + 12 \qquad x - 8 + 8 = 0 + 8$$

$$x = 12 \qquad\qquad x = 8$$

Evaluation: The sum of $12 + 8 = 20$. The sum of the reciprocals is $\dfrac{1}{12} + \dfrac{1}{8} = \dfrac{2}{24} + \dfrac{3}{24} = \dfrac{5}{24}$.

TRY THESE

1. One number is 3 more than another number, and the product of the two numbers is 54. Find the numbers.

2. If the product of two consecutive even numbers is 168, find the numbers.

3. If 8 is subtracted from the square of a number, the answer is 28. Find the numbers.

4. Mike can paint a room in 16 minutes less time than Ike. If they both paint the room at the same time, it will take them 15 minutes. How long does it take each one to paint the room individually?

5. One number is 4 more than another number. If the square of the smaller number is 2 less than three times the larger number, find the numbers.

6. Beverly is two years older than Mary. If the product of their ages is 48, find each one's age.

7. Two square plots of land contain 74 square feet. If the side of one plot is 2 feet longer than the side of the other plot, find the dimensions of both plots. (The formula for the area of a square is $A = s^2$.)

8. The sum of a number and its reciprocal is $7\frac{1}{7}$. Find the number.

9. Two workers can assemble a trailer in six hours. If it takes the second worker nine hours longer than the first worker to assemble the trailer, how long will it take each worker to do the job if they work alone?

10. If the sum of the squares of two consecutive numbers is 85, find the numbers.

✔ SOLUTIONS

1. Let x = one number and $x + 3$ = the other number.

$$x(x + 3) = 54$$
$$x^2 + 3x = 54$$
$$x^2 + 3x - 54 = 54 - 54$$
$$x^2 + 3x - 54 = 0$$
$$(x + 9)(x - 6) = 0$$

$$x + 9 = 0 \qquad\qquad x - 6 = 0$$
$$x + 9 - 9 = 0 - 9 \qquad x + 6 - 6 = 0 + 6$$
$$x = -9 \qquad\qquad x = 6$$
$$x + 3 = -9 + 3 = -6 \qquad x + 3 = 6 + 3 = 9$$

The answers are –6 and –9, and 6 and 9.

2. Let $x =$ one number and $x + 2 =$ the other number.

$$x(x + 2) = 168$$
$$x^2 + 2x = 168$$
$$x^2 + 2x - 168 = 168 - 168$$
$$x^2 + 2x - 168 = 0$$
$$(x + 14)(x - 12) = 0$$

$$x + 14 = 0 \qquad\qquad x - 12 = 0$$
$$x + 14 - 14 = 0 - 14 \qquad\qquad x - 12 + 12 = 0 + 12$$
$$x = -14 \qquad\qquad x = 12$$
$$x + 2 = -14 + 2 = -12 \qquad\qquad x + 2 = 12 + 2 = 14$$

The answers are –14 and –12, and 14 and 12.

3. Let $x =$ the number.

$$x^2 - 8 = 28$$
$$x^2 - 8 - 28 = 28 - 28$$
$$x^2 - 36 = 0$$
$$(x - 6)(x + 6) = 0$$

$$x - 6 = 0 \qquad\qquad x + 6 = 0$$
$$x - 6 + 6 = 0 + 6 \qquad\qquad x + 6 - 6 = 0 - 6$$
$$x = 6 \qquad\qquad x = -6$$

The answers are 6 and –6.

4. Let $x =$ the time it takes Mike to paint the room and $x + 16 =$ the time it takes Ike to paint the room.

$$\frac{15}{x} + \frac{15}{x + 16} = 1$$

$$\text{LCD} = x(x + 16)$$

$$\cancel{x}^{1}(x + 16) \cdot \frac{15}{\cancel{x}^{1}} + x(\cancel{x + 16})^{1} \cdot \frac{15}{\cancel{x + 16}^{1}} = x(x + 16)$$

$$15(x + 16) + 15x = x(x + 16)$$

$$15x + 240 + 15x = x^2 + 16x$$

$$30x + 240 = x^2 + 16x$$

$$30x - 30x + 240 = x^2 + 16x - 30x$$

$$240 = x^2 - 14x$$

$$240 - 240 = x^2 - 14x - 240$$

$$0 = x^2 - 14x - 240$$

$$0 = (x - 24)(x + 10)$$

$x - 24 = 0$	$x + 10 = 0$
$x - 24 = 0 + 24$	$x + 10 - 10 = 0 - 10$
$x = 24$	$x = -10$
$x + 16 = 24 + 16 = 40$	Ignore $x = -10$ since it is meaningless.

It takes Mike 24 minutes to paint the room and Ike 40 minutes to paint the room.

5. **Let $x =$ the smaller number and $x + 4 =$ the larger number.**

$$x^2 + 2 = 3(x + 4)$$

$$x^2 + 2 = 3x + 12$$

$$x^2 + 2 - 3x = 3x - 3x + 12$$

$$x^2 - 3x + 2 = 12$$

$$x^2 - 3x + 2 - 12 = 12 - 12$$

$$x^2 - 3x - 10 = 0$$

$$(x - 5)(x + 2) = 0$$

$x - 5 = 0$	$x + 2 = 0$
$x - 5 + 5 = 0 + 5$	$x + 2 - 2 = 0 - 2$
$x = 5$	$x = -2$
$x + 4 = 5 + 4 = 9$	$x + 4 = -2 + 4 = 2$

The answers are 5 and 9 and −2 and 2.

6. **Let $x =$ Mary's age and $x + 2 =$ Beverly's age.**

$$x(x + 2) = 48$$

$$x^2 + 2x = 48$$

$$x^2 + 2x - 48 = 48 - 48$$

$$x^2 + 2x - 48 = 0$$

$$(x + 8)(x - 6) = 0$$

$$x + 8 = 0 \qquad\qquad x - 6 = 0$$

$$x + 8 - 8 = 0 - 8 \qquad x - 6 + 6 = 0 + 6$$

$$x = -8 \qquad\qquad x = 6$$

Ignore this result. $x + 2 = 6 + 2 = 8$

Hence, Mary is 6 years old and Beverly is 8 years old.

7. Let x be the length of the side of one plot and x^2 be the area. Let $(x + 2)$ be the length of the side of the other plot and $(x + 2)^2$ be the area.

$$x^2 + (x + 2)^2 = 74$$

$$x^2 + x^2 + 4x + 4 = 74$$

$$2x^2 + 4x + 4 - 74 = 74 - 74$$

$$2x^2 + 4x - 70 = 0$$

$$\frac{\cancel{2}^1 x^2}{\cancel{2}^1} + \frac{\cancel{4}^2 x}{\cancel{2}^1} - \frac{\cancel{70}^{35}}{\cancel{2}^1} = 0$$

$$x^2 + 2x - 35 = 0$$

$$(x + 7)(x - 5) = 0$$

$$x + 7 = 0 \qquad\qquad x - 5 = 0$$

$$x + 7 - 7 = 0 - 7 \qquad x - 5 + 5 = 0 + 5$$

$$x = -7 \qquad\qquad x = 5$$

Ignore this result. $x + 2 = 5 + 2 = 7$

Hence, the side of one plot is 7 feet and the side of the other plot is 5 feet.

8. Let $x =$ the number and $\dfrac{1}{x} =$ the reciprocal of the number.

$$x + \frac{1}{x} = 7\frac{1}{7}$$

$$x + \frac{1}{x} = \frac{50}{7}$$

$$7x \cdot x + 7\cancel{x}^1 \cdot \frac{1}{\cancel{x}^1} = \cancel{7}^1 x \cdot \frac{50}{\cancel{7}^1}$$

$$7x^2 + 7 = 50x$$

$$7x^2 + 7 - 50x = 50x - 50x$$

$$7x^2 - 50x + 7 = 0$$

$$(7x - 1)(x - 7) = 0$$

$$7x - 1 = 0 \qquad\qquad x - 7 = 0$$

$$7x - 1 + 1 = 0 + 1 \qquad x - 7 + 7 = 0 + 7$$

$$7x = 1 \qquad\qquad x = 7$$

$$\frac{\cancel{7}^{1} x}{\cancel{7}^{1}} = \frac{1}{7}$$

$$x = \frac{1}{7}$$

Hence, the number is 7 or $\dfrac{1}{7}$.

9. Let $x =$ the time the first worker takes to do the job and $x + 9 =$ the time the second worker takes to do the job.

$$\frac{6}{x} + \frac{6}{x + 9} = 1$$

$$\cancel{x}^{1}(x + 9) \cdot \frac{6}{\cancel{x}^{1}} + x(\cancel{x + 9})^{1} \cdot \frac{6}{\cancel{x + 9}^{1}} = x(x + 9)$$

$$6(x + 9) + 6x = x(x + 9)$$

$$6x + 54 + 6x = x^2 + 9x$$

$$12x + 54 = x^2 + 9x$$

$$12x - 12x + 54 = x^2 + 9x - 12x$$

$$54 = x^2 - 3x$$

$$54 - 54 = x^2 - 3x - 54$$

$$0 = x^2 - 3x - 54$$

$$0 = (x - 9)(x + 6)$$

$$x - 9 = 0 \qquad\qquad x + 6 = 0$$

$$x - 9 + 9 = 0 + 9 \qquad x + 6 - 6 = 0 - 6$$

$$x = 9 \qquad\qquad x = -6$$

$$x + 9 = 9 + 9 = 18 \qquad \text{Ignore this result.}$$

Hence, it will take one worker 9 hours to assemble the trailer and the other worker 18 hours to do the job.

10. Let x = the first number and $x + 1$ = the next number.

$$x^2 + (x + 1)^2 = 85$$

$$x^2 + x^2 + 2x + 1 = 85$$

$$2x^2 + 2x + 1 = 85$$

$$2x^2 + 2x + 1 - 85 = 85 - 85$$

$$2x^2 + 2x - 84 = 0$$

$$\frac{\cancel{2}^1 x^2}{\cancel{2}^1} + \frac{\cancel{2}^1 x}{\cancel{2}^1} - \frac{\cancel{84}^{42}}{\cancel{2}^1} = 0$$

$$x^2 + x - 42 = 0$$

$$(x + 7)(x - 6) = 0$$

$x + 7 = 0$	$x - 6 = 0$
$x + 7 - 7 = 0 - 7$	$x - 6 + 6 = 0 + 6$
$x = -7$	$x = 6$
$x + 1 = -7 + 1 = -6$	$x + 1 = 6 + 1 = 7$

Hence, the answers are −7 and −6 and 7 and 6.

Summary

In this chapter, you learned how to solve word problems using a quadratic equation. Many of these equations can be solved by factoring. It is important to realize that many quadratic equations cannot be solved by factoring so the quadratic formula can be used. This formula can be found in an algebra book.

QUIZ

1. If the product of two positive consecutive odd number is 323, find the larger one.

   A. 19
   B. 21
   C. 15
   D. 17

2. A person has two square foundations for two sheds. The total of the areas of both foundations is 73 square feet. If the side of one foundation is 5 feet longer than the side of the other one, find the length of the smaller foundation. (Use $A = s^2$.)

   A. 6 feet
   B. 2 feet
   C. 5 feet
   D. 3 feet

3. If the length of a rectangle is 5 inches longer than its width and the area of the rectangle is 24 square feet, find the length of the rectangle. (Use $A = lw$.)

   A. 8 feet
   B. 6 feet
   C. 4 feet
   D. 3 feet

4. If the side of a square is increased by 3 inches, the area of the square is 324 square inches. If the side of the same square is decreased by 3 inches, the area of the square is 144 square inches. Find the measure of the side of the square. (Use $A = s^2$.)

   A. 7 inches
   B. 15 inches
   C. 18 inches
   D. 12 inches

5. If Dave is four years older than Jim and the product of their ages is 117, how old is Dave?

   A. 9
   B. 11
   C. 13
   D. 15

6. If the sum of square of two consecutive numbers is 61, find the smaller number.

   A. 6
   B. 8
   C. 7
   D. 5

7. One side of a square is three inches longer than the side of another square. If the sum of their areas is 185 square inches, find the length of the side of the longer-sided square. (Use $A = s^2$.)

   A. 11 square inches
   B. 9 square inches
   C. 12 square inches
   D. 7 square inches

8. Bret is 4 years older than Sam. If Sam's age is squared, the result is 26 more than Bret's age. Find Sam's age.

   A. 5
   B. 6
   C. 8
   D. 3

9. If the difference between a number and its reciprocal is $7\dfrac{7}{8}$, find the whole number.

   A. 7
   B. 9
   C. 8
   D. 6

10. Two workers working together can clean a small office building in 4.8 hours. One worker can do it in 4 hours less time than the other. Find the time it would take the slower worker to clean the building if he works by himself.

    A. 8 hours
    B. 10 hours
    C. 12 hours
    D. 14 hours

chapter 12

Solving Word Problems in Geometry, Probability, and Statistics

This chapter explains how to solve word problems in geometry, probability, and statistics. These problems are only a sample of the types of problems that you will find in these courses, since there are entire books written on these subjects.

CHAPTER OBJECTIVES

In this chapter, you will learn how to

- Solve word problems in geometry
- Solve word problems in probability
- Solve word problems in statistics

Solving Geometry Problems

Although the word problems in geometry are for the most part different from those in algebra, many problems in geometry require algebra to solve them. Since it is not possible to show all the different types of problems that you will find in geometry, a few of them will be explained here so that you can reach a basic understanding of how to use algebra to solve some of the problems found in geometry.

Each problem is based on a geometric principle or rule. The principles will be given here in each problem.

EXAMPLE

Find the measure of each angle of a triangle if the measure of the second angle is twice as large as the measure of the first angle and the third angle is three times the measure of the first angle.

Geometric principle: The sum of the measures of the angles of a triangle is 180°.

SOLUTION

Goal: You are being asked to find the measures of the three angles of a triangle.

Strategy: Let x = the measure of the first angle

$2x$ = the measure of the second angle

$3x$ = the measure of the third angle

See Figure 12-1.

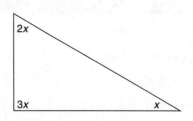

FIGURE 12-1

Since the sum of the measures of the angles of a triangle is 180°, the equation is

$$x + 2x + 3x = 180°$$

Implementation: Solve the equation:

$$x + 2x + 3x = 180°$$

$$6x = 180°$$

$$\frac{\cancel{6}^{1} x}{\cancel{6}^{1}} = \frac{180°}{6x}$$

$$x = 30°$$

$$2x = 2 \cdot 30° = 60°$$

$$3x = 3 \cdot 30° = 90°$$

Hence, the measures of the angles are 30°, 60°, and 90°.

Evaluation: Check that the sum of the angles is 180°.

$$30° + 60° + 90° = 180°$$

EXAMPLE

If the length of a rectangle is three times its width and the perimeter of the rectangle is 104 inches, find the measures of its length and width.

Geometric principle: The perimeter of a rectangle is $P = 2l + 2w$.

SOLUTION

Goal: You are being asked to find the length and width of a rectangle.

Strategy: Let x = the width of the rectangle and $3x$ = the length of the rectangle.

See Figure 12-2.

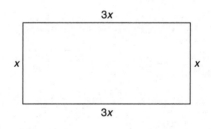

FIGURE 12-2

Since the formula for the perimeter of a rectangle is $P = 2l + 2w$, the equation is

$$2(3x) + 2(x) = 104.$$

Implementation: Solve the equation:

$$2(3x) + (2x) = 104$$
$$6x + 2x = 104$$
$$8x = 104$$

$$\frac{\cancel{8}^1 x}{\cancel{8}^1} = \frac{104}{8}$$

$$x = 13$$
$$3x = 3 \cdot 13 = 39$$

Hence, the length is 39 inches and the width is 13 inches.

Evaluation: Use the formula for perimeter and check that it is 104 inches.

$$P = 2l + 2w$$
$$P = 2(39) + 2(13)$$
$$= 78 + 26$$
$$= 104 \text{ inches}$$

EXAMPLE

The base of a triangle is 8 inches longer than its height. If the area of the triangle is 10 square inches, find the base and height of the triangle.

Geometric principle: The area of a triangle is $A = \dfrac{1}{2} bh$.

SOLUTION

Goal: You are being asked to find the measures of the base and the height.

Strategy: Let x = the measure of the height and $x + 8$ = the measure of the base.

See Figure 12-3.

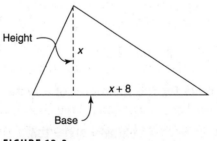

FIGURE 12-3

Since $A = \frac{1}{2}bh$, the equation is $10 = \frac{1}{2}(x + 8)x$.

Implementation: Solve the equation:

$$10 = \frac{1}{2}(x + 8)x$$

$$20 = x^2 + 8x$$

$$20 - 20 = x^2 + 8x - 20$$

$$0 = x^2 + 8x - 20$$

$$0 = (x + 10)(x - 2)$$

$$x + 10 = 0 \qquad\qquad x - 2 = 0$$

$$x + 10 - 10 = 0 - 10 \qquad\qquad x - 2 + 2 = 0$$

$$x = -10 \qquad\qquad x = 2$$

In this case, we ignore $x = -10$ since a height cannot be a negative number. The base is $x + 8 = 2 + 8 = 10$ inches. Hence, the height is 2 inches and the base is 10 inches.

Evaluation: Find the area and check that it is 10 square inches.

$$A = \frac{1}{2}bh$$

$$= \frac{1}{2}(10)(2)$$

$$= 10 \text{ square inches}$$

TRY THESE

1. If an angle exceeds its complement by 32°, find its measure. *Geometric principle:* Complementary angles are adjacent angles whose sum is 90°.

2. The area of a rectangle is 80 square inches. The length is 16 inches longer than the width. Find the dimensions of the rectangle. *Geometric principle:* The area of a rectangle is $A = lw$.

3. The perimeter of a rectangle is 64 inches and the length is three times the width. Find its dimensions. *Geometric principle:* The perimeter of a rectangle is $P = 2l + 2w$.

4. The perimeter of a rectangle is 76 inches. If the length is 14 inches more than twice the width, find its dimensions. *Geometric principle:* The perimeter of a rectangle is $P = 2l + 2w$.

5. If the side of a large square is four times as long as the side of a smaller square and the area of the large square is 375 square inches larger than the area of the smaller square, find the length of the side of the smaller square. *Geometric principle:* The area of a square is $A = s^2$.

6. The sum of the measures of the angles of a triangle is 180°. If the measure of the second angle is twice as large as the measure of the first angle and the measure of the third angle is 20° more than the measure of the second angle, find the measures of the angles.

7. The base of a triangle is 11 feet longer than its height. If its area is 30 square feet, find the measures of the base and height. *Geometric principle:* The area of a triangle is $A = \dfrac{1}{2}bh$.

8. If two sides of a triangle are equal in length and the third side is 10 inches shorter than the length of one of the equal sides, find the length of the sides if the perimeter is 29 inches. *Geometric principle:* The perimeter of a triangle is equal to the sum of the lengths of its sides.

9. If the area of a circle is 314 square inches, find the radius. *Geometric principle:* The area of a circle is $A = 3.14r^2$.

10. If one angle of a triangle is 42° more than twice another angle, and the third angle is equal to the sum of the first two angles, find the measure of each angle. *Geometric principle:* The sum of the measure of the angles of a triangle is 180°.

☑ SOLUTIONS

1. Let x = the measure of one angle and $x + 32°$ = the measure of the larger angle.

$$x + x + 32° = 90$$
$$2x + 32 = 90$$
$$2x + 32 - 32 = 90 - 32$$
$$2x = 58$$

$$\frac{\cancel{2}^1 x}{\cancel{2}^1} = \frac{58}{2}$$

$$x = 29°$$
$$x + 32° = 29° + 32° = 61°$$

The measures of the angles are 29° and 61°.

2. Let x = the width of the rectangle and $x + 16$ = the length of the rectangle.

$$A = lw$$
$$80 = (x + 16)x$$
$$80 = x^2 + 16x$$
$$0 = x^2 + 16x - 80$$
$$0 = (x + 20)(x - 4)$$

$x + 20 = 0$	$x - 4 = 0$
$x + 20 - 20 = 0 - 20$	$x - 4 + 4 = 0 + 4$
$x = -20$	$x = 4$
Ignore this answer.	$x + 6 = 4 + 16 = 20$

The length of the rectangle is 20 inches and the width is 4 inches.

3. Let x = the width and $3x$ = the length.

$$P = 2l + 2w$$
$$64 = 2(3x) + 2x$$
$$64 = 6x + 2x$$
$$64 = 8x$$
$$\frac{64}{8} = \frac{\cancel{8}^1 x}{\cancel{8}^1}$$
$$8 = x$$
$$3x = 3 \cdot 8 = 24$$

The length is 24 inches and the width is 8 inches.

4. Let x = the width of the rectangle and $2x + 14$ = the length of the rectangle.

$$P = 2l + 2w$$
$$76 = 2(2x + 14) + 2x$$
$$76 = 4x + 28 + 2x$$
$$76 = 6x + 28$$
$$76 - 28 = 6x + 28 - 28$$
$$48 = 6x$$

$$\frac{48}{6} = \frac{\cancel{1}^1 x}{\cancel{1}^1}$$

$$8 = x$$

$$2x + 14 = 2(8) + 14 = 16 + 14 = 30$$

The length is 30 inches and the width is 8 inches.

5. Let x = the length of the side of the smaller square and $4x$ = the length of the larger square.

$$(4x)^2 - x^2 = 375$$

$$16x^2 - x^2 = 375$$

$$15x^2 = 375$$

$$\frac{\cancel{15}^1 x^2}{\cancel{15}^1} = \frac{375}{15}$$

$$x^2 = 25$$

$$x = \sqrt{25} = 5$$

The length of the side of the smaller square is 5 inches.

6. Let x = the measure of one angle

$2x$ = the measure of the second angle

$2x + 20$ = the measure of the third angle

$$180 = x + 2x + 2x + 20$$

$$180 = 5x + 20$$

$$180 - 20 = 5x + 20 - 20$$

$$160 = 5x$$

$$\frac{160}{5} = \frac{\cancel{5}^1 x}{\cancel{5}^1}$$

$$32° = x$$

$$2x = 2(32) = 64°$$

$$2x + 20 = 2(32) + 20 = 64 + 20 = 84°$$

The measures of the three angles are 32°, 64°, and 84°.

7. Let x = the measure of the height and $x + 11$ = the measure of the base.

$$A = \frac{1}{2}bh$$

$$30 = \frac{1}{2}(x + 11)x$$

$$30 \cdot 2 = \frac{\cancel{2}^{1}}{1} \cdot \frac{1}{\cancel{2}^{1}}(x + 11)x$$

$$60 = x^2 + 11x$$

$$60 - 60 = x^2 + 11x - 60$$

$$0 = x^2 + 11x - 60$$

$$0 = (x + 15)(x - 4)$$

$x + 15 = 0$	$x - 4 = 0$
$x + 15 - 15 = 0 - 15$	$x - 4 + 4 = 0 + 4$
$x = -15$	$x = 4$
Ignore this result.	$x + 11 = 4 + 11 = 15$

The height is 4 feet and the base is 15 feet.

8. Let x = the length of one of the two equal sides and $x - 10$ = the length of the third side.

$$x + x + x - 10 = 29$$

$$3x - 10 = 29$$

$$3x - 10 + 10 = 29 + 10$$

$$3x = 39$$

$$\frac{\cancel{3}^{1}x}{\cancel{3}^{1}} = \frac{39}{3}$$

$$x = 13$$

$$x - 10 = 13 - 10 = 3$$

The lengths of the sides are 13 inches, 13 inches, and 3 inches.

9. Let x = the measure of the radius.

$$A = 3.14r^2$$

$$314 = 3.14x^2$$

$$\frac{314}{3.14} = \frac{\cancel{3.14}^1 x^2}{\cancel{3.14}^1}$$

$$100 = x^2$$

$$\sqrt{100} = x$$

$$10 = x$$

The radius is 10 inches.

10. Let x = the measure of one angle

$2x + 42$ = the measure of the second angle

$x + 2x + 42$ = the measure of the third angle

$$x + 2x + 42 + x + 2x + 42 = 180$$

$$6x + 84 = 180$$

$$6x + 84 - 84 = 180 - 84$$

$$6x = 96$$

$$\frac{\cancel{6}^1 x}{\cancel{6}^1} = \frac{96}{6}$$

$$x = 16°$$

$$2x + 42 = 2(16) + 42 = 74°$$

$$16° + 74° = 90°$$

The measures of the angles are 16°, 74°, and 90°.

In this section, you learned how to solve some kinds of word problems in geometry. Many of these problems use geometric formulas and some basic algebra.

Solving Probability Problems

Probability deals with chance events, such as card games, slot machines, and lotteries as well as insurance, investments, and weather forecasting. A *probability experiment* is a chance process that leads to well-defined outcomes. For example, when a die (singular for dice) is rolled, there are six possible well-defined outcomes.

They are

$$1, 2, 3, 4, 5, 6$$

When a coin is flipped, there are two possible well-defined outcomes. They are

$$\text{heads, tails}$$

The set of all possible outcomes of a probability experiment is called the *sample space*. Each outcome in a sample space, unless otherwise noted, is considered equally likely that is, it has the same chance of occurring. An *event* can consist of outcomes in the sample space. The basic definition of the probability of an event is

$$P(E) = \frac{\text{number of outcomes in event } E}{\text{total number of outcomes in the sample space}}$$

The strategy when determining the probability of an event is

1. Find the number of outcomes in event E.

2. Find the number of outcomes in the sample space.

3. Divide the first number by the second number to get a decimal or reduce the fraction if a fraction answer is desired.

EXAMPLE

A die is rolled; find the probability of getting an even number.

SOLUTION

Goal: You are being asked to find the probability of getting an even number.

Strategy: When a die is rolled, there are six outcomes in the sample space, and there are three outcomes in the event—that is, there are three even numbers: 2, 4, and 6.

Implementation: $P(\text{even number}) = \dfrac{3}{6} = \dfrac{1}{2}$ or 0.5.

Evaluation: Since 2, 4, and 6 are half of the numbers in the sample space, the probability is correct.

When two coins are tossed, the sample space is

HH, HT, TH, TT

EXAMPLE

Two coins are tossed. Find the probability of getting two heads.

SOLUTION

Goal: **You are being asked to find the probability of getting two heads.**

Strategy: **There are four outcomes in the sample space, and there is only one way two heads can occur.**

Implementation: $P = \text{(two heads)} = \dfrac{1}{4}.$

Evaluation: **Looking at the sample space, it is obvious that the probability of one choice from four outcomes is $\dfrac{1}{4}$.**

When two dice are rolled, each die can have one of six outcomes. Therefore, there are $6 \times 6 = 36$ outcomes in the sample space. The outcomes can be arranged in ordered pairs such that the first number is the number of spots on the first die, and the second number in the pair is the number of spots on the second die. For example, the ordered pair (2, 4) means a 2 came up on the first die and a 4 came up on the second die. Also, the sum of the numbers for this outcome is $2 + 4 = 6$.

The sample space for two dice is shown next:

$$
\begin{array}{cccccc}
(1, 1) & (1, 2) & (1, 3) & (1, 4) & (1, 5) & (1, 6) \\
(2, 1) & (2, 2) & (2, 3) & (2, 4) & (2, 5) & (2, 6) \\
(3, 1) & (3, 2) & (3, 3) & (3, 4) & (3, 5) & (3, 6) \\
(4, 1) & (4, 2) & (4, 3) & (4, 4) & (4, 5) & (4, 6) \\
(5, 1) & (5, 2) & (5, 3) & (5, 4) & (5, 5) & (5, 6) \\
(6, 1) & (6, 2) & (6, 3) & (6, 4) & (6, 5) & (6, 6)
\end{array}
$$

EXAMPLE

Two dice are rolled; find the probability of getting a sum of 6.

SOLUTION

Goal: **You are being asked to find the probability of getting a sum of 6.**

Strategy: There are 36 outcomes in the sample space and five ways to get a sum of six. They are (1, 5), (2, 4), (3, 3), (4, 2), and (5, 1).

Implementation: $P(\text{sum of 6}) = \dfrac{5}{36}$.

Evaluation: Use the sample space to verify your answer.

EXAMPLE

Two dice are rolled; find the probability of getting a sum greater than 9.

SOLUTION

Goal: You are being asked to find the probability of getting a sum greater than 9.

Strategy: A sum greater than 9 means a sum of 10, 11, or 12. They are (4, 6), (5, 5), (6, 4), (5, 6), (6, 5) and (6, 6). Hence, there are six ways to get a sum greater than 9, and there are 36 outcomes in the sample space.

Implementation: $P(\text{sum greater than 9}) = \dfrac{6}{36} = \dfrac{1}{6}$.

Evaluation: Use the sample space to verify the answer.

Probability problems also use ordinary playing cards. In a deck of cards, there are 52 cards consisting of four suits: hearts and diamonds, which are red, and spades and clubs, which are black. In addition, there are 13 cards in each suit, ace through ten and a jack, a queen, and a king (called face cards). See Figure 12-4.

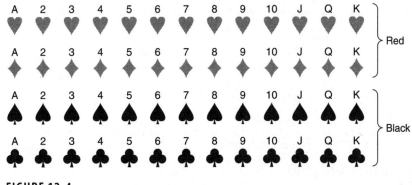

FIGURE 12-4

EXAMPLE

A card is drawn from a deck. Find the probability that it is a king.

SOLUTION

Goal: You are being asked to find the probability that the selected card is a king.

Strategy: There are 52 outcomes in the sample space, and four of them are kings.

Implementation: $P(4 \text{ kings}) = \dfrac{4}{52} = \dfrac{1}{13}$.

Evaluation: Use the sample space to verify the answer.

EXAMPLE

A card is selected from a deck; find the probability that it is a diamond.

SOLUTION

Goal: You are being asked to find the probability of selecting a diamond.

Strategy: There are 13 diamonds in a deck of 52 cards.

Implementation: $P(\text{diamond}) = \dfrac{13}{52} = \dfrac{1}{4}$.

Evaluation: Use the sample space to verify your answer.

The examples shown previously are examples of what is called *classical probability*. The next examples are from another area of probability called *empirical probability*. Empirical probability uses *frequency distributions*. Suppose that a bag of mixed candy contained six caramels, three peppermints, seven chocolates, and nine coconut creams. The sample space can be represented using a frequency distribution as shown.

Candy	Frequency
Caramels	6
Peppermints	3
Chocolates	7
Coconut creams	9
Total	25

This distribution can be used to solve probability problems.

EXAMPLE

Suppose a person selects a piece of candy from the bag; find the probability that it is a caramel.

SOLUTION

Goal: You are being asked to find the probability that the piece of candy is a caramel.

Strategy: There are 6 caramels and a total of 25 pieces of candy, so the probability formula can be used.

Implementation: $P(\text{caramel}) = \dfrac{6}{25}$.

Evaluation: The answer can be verified by looking at the frequency distribution.

EXAMPLE

Using the same bag of candy, find the probability that a person selects a peppermint or a chocolate.

SOLUTION

Goal: You are being asked to find the probability that the piece of candy is a peppermint or a chocolate.

Strategy: There are 25 pieces of candy and there are 3 peppermints and 7 chocolates.

Implementation: $P(\text{peppermint or chocolate}) = \dfrac{3+7}{25} = \dfrac{10}{25} = \dfrac{2}{5}$.

Evaluation: You can verify the answer by looking at the frequency distribution.

EXAMPLE

In a classroom, there are 20 juniors and 8 seniors. If a student is selected at random to read a passage, find the probability that the student is a senior.

SOLUTION

Goal: You are being asked to find the probability that the student is a senior.

Strategy: There are a total of 28 students in the class and 8 are seniors.

Implementation: $P(\text{senior}) = \dfrac{8}{28} = \dfrac{2}{7}$.

Evaluation: The answer can be verified by looking at the problem.

There are four basic rules for probability problems:

Rule 1: *The probability of any event is a number from zero to 1.* This means that an answer in a probability problem can never be less than zero (i.e., negative) or greater than 1.

Rule 2: *If the probability of an event is zero, the event cannot occur.* For example, if you roll a single die, find the probability of getting a 9. Since a 9 cannot occur when you roll a single die (a regular die has only six sides and six numbers), $P(9) = 0/6 = 0$.

Rule 3: *If the probability of an event is 1, the event is certain to occur.* For example, if you take out all of the black cards from a deck of 52 cards, you have 26 red cards left. Now if you select one card, the probability that it will be red will be $P(\text{red card}) = 26/26 = 1$. In other words, a red card is certain to occur.

Rule 4: *The sum of the probabilities of all the events in the sample space will be 1.* In other words, if you take each event in the sample space, find its probability, and add all the values, you will always get 1. For example, if you roll a single die, the probability of getting each number is 1/6, and since there are six possible outcomes, the sum of these probabilities will be $1/6 + 1/6 + 1/6 + 1/6 + 1/6 + 1/6 = 6/6 = 1$.

Another important aspect of probability is that the closer the probability of an event is to 1, the more likely the event will occur. On the other hand, the closer the probability of an event is to zero, the less likely the event will occur.

Sometimes in probability problems, you will be asked to find the probability that one event *or* another event will occur. The word "or" in this case means to add the individual probabilities. For example, if you draw one card from the deck, the probability that it will be king or a queen will be $4/52 + 4/52 = 8/52 = 2/13$ since the individual probabilities are 4/52 and 4/52. There are four kings and four queens. Notice that these two events cannot occur at the same time. They are called *mutually exclusive events*. Now, what if you draw a single card from a deck and you are asked to find the probability of getting a 7 or a club? In this case, there are four 7s, so $P(7) = 4/52$, and there are 13 clubs, so $P(\text{club}) = 13/52$. If you add the probabilities of getting a 7 or a club, you will get $4/52 + 13/52 = 17/52$. This is the wrong answer, since the 7 of clubs was counted twice. In other words, these two events are *not* mutually exclusive. When two events are *not* mutually exclusive, you must subtract the probability that the events occur at the same time. So P(7 of clubs) $= 1/52$. Hence, P(7 or club) $= 4/52 + 13/52 - 1/52 = 16/52 = 4/13$.

The two rules are summarized as follows:

When two outcomes are mutually exclusive, $P(A \text{ or } B) = P(A) + P(B)$.

When two outcomes are not mutually exclusive, $P(A \text{ or } B) = P(A) + P(B) - P(A \text{ and } B)$, where $P(A \text{ and } B)$ is the probability that the outcomes occur at the same time.

EXAMPLE

Draw a card from a deck. Find the probability that it is a red card or an ace.

SOLUTION

Goal: **You are being asked to find the probability that the card selected is a red card or an ace.**

Strategy: **There are 26 red cards and four aces; however, two of the aces are red.**

Implementation:

$$P(\text{red card or ace}) = P(\text{red card}) + P(\text{ace}) - P(\text{red aces}) = \frac{26}{52} + \frac{4}{52} - \frac{2}{52} = \frac{28}{52} = \frac{7}{13}.$$

Evaluation: **You can look at the sample space and count the red cards and the two aces that are not red. You get 28 different cards. Hence, the answer is $\frac{28}{52}$ or $\frac{7}{13}$.**

Another type of probability problem happens when you perform the probability experiment more than once. For example, suppose you roll a die three times and you are asked to find the probability for a certain outcome such as getting three 6s. Other examples might be drawing two cards from a deck or flipping five coins.

In these types of problems, you have to determine whether or not the outcome of the first time you perform the experiment affects or changes the probability of the outcome of the second time you do the experiment.

For example, when you flip a coin twice or roll a die three times, the outcome of the first time does not affect the outcome of the second time you do the experiment. When you flip a coin, the probability of getting a head each time is always one-half. No matter how many times you roll a die, the probability of getting a 3 will always be 1/6. In these cases, the outcomes are said to be *independent* of each other.

When you draw two cards from a deck and replace the first card before you select the second card, the outcomes are independent, but if you do not replace the card before selecting the second card, the probability changes. These outcomes are said to be *dependent*.

These two rules can be summarized as follows:

When two events are independent, $P(A \text{ and } B) = P(A) \times P(B)$.

When two events are dependent, $P(A \text{ and } B) = P(A) \times P(B \text{ given that } A \text{ has occurred})$.

EXAMPLE

Draw two cards from a deck without replacement. Find the probability of getting two kings.

SOLUTION

Goal: You are being asked to find the probability of getting two kings when two cards are drawn from a deck without replacing the first card after it is drawn.

Strategy: There are four kings in a deck of 52 cards so $P(K) = \dfrac{4}{52}$. Now if a king occurs on the first draw, there are three kings left and 51 cards remaining in the deck. So $P(K \text{ on second draw}) = \dfrac{3}{51}$.

Implementation: Apply the rule $P(A \text{ and } B) = P(A) \cdot P(B \text{ given that } A \text{ has already occurred}) = \dfrac{4}{52} \times \dfrac{3}{51} = \dfrac{\overset{1}{\cancel{4}}}{\underset{13}{\cancel{52}}} \times \dfrac{\overset{1}{\cancel{3}}}{\underset{17}{\cancel{51}}} = \dfrac{1}{221}$.

Evaluation: These types of problems are difficult to evaluate, so use a little common sense or reasoning and check your arithmetic.

Notice that if the first card is replaced after the first draw, the outcomes are independent and $P(2 \text{ kings}) = \dfrac{4}{52} \times \dfrac{4}{52} = \dfrac{\overset{1}{\cancel{4}}}{\underset{13}{\cancel{52}}} \times \dfrac{\overset{1}{\cancel{4}}}{\underset{13}{\cancel{52}}} = \dfrac{1}{169}$.

TRY THESE

1. A single die is rolled once; find the probability of getting

   a. a 3

   b. a number greater than 2

   c. a number less than 7

   d. a number greater than 6

2. Two dice are rolled; find the probability of getting

   a. a sum of 9

   b. doubles

   c. a sum greater than 10

   d. a sum less than 4

3. A card is drawn from a deck; find the probability of getting

   a. the 7 of spades

   b. a jack

   c. a club

   d. a heart or a club

   e. a red card

4. A couple has three children; find the probability that the children are

   a. all girls

   b. all boys or all girls

   c. exactly two girls and one boy

5. Two dice are rolled; find the probability of getting a sum of 8 or 10.

6. In a cooler there are nine cans of cola and six cans of cherry soda. If a person selects a can of soda without looking at it, find the probability that it is a can of cola.

7. Two dice are rolled; find the probability of getting a sum greater than 8 or doubles.

8. A box contains three orange balls, two blue balls, and one red ball. If two balls are selected without replacement, find the probability of getting two orange balls.

9. A die is rolled three times; find the probability of getting an even number all three times.

10. A die is rolled twice. Find the probability of getting the same number twice.

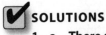**SOLUTIONS**

1. a. There are six outcomes in the sample space and one outcome is a 3; therefore, $P(3) = \dfrac{1}{6}$.

    b. There are six outcomes in the sample space and there are four outcomes that are greater than 2; that is, 3, 4, 5, and 6; hence, $P(\text{a number greater than 2}) = \dfrac{4}{6} = \dfrac{2}{3}$.

    c. There are six outcomes in the sample space and six numbers less than 7; hence, $P(\text{a number less than 7}) = \dfrac{6}{6} = 1$.

    d. There are six outcomes in the sample space and no numbers are greater than 6; hence, $P(\text{a number greater than 6}) = \dfrac{0}{6} = 0$.

2. a. There are 36 outcomes in the sample space and there are four ways to get a sum of 9: (3, 6), (4, 5), (5, 4), and (6, 3); hence, $P(\text{sum of 9}) = \dfrac{4}{36} = \dfrac{1}{9}$.

    b. There are 36 outcomes in the sample space and six ways to get doubles: (1, 1), (2, 2), (3, 3), (4, 4), (5, 5), and (6, 6); hence $P(\text{doubles}) = \dfrac{6}{36} = \dfrac{1}{6}$.

    c. There are 36 outcomes in the sample space and two sums greater than 10—that is, a sum of 11, or 12: (5, 6), (6, 5), and (6, 6). Hence, $P(\text{sum greater than 10}) = \dfrac{3}{36} = \dfrac{1}{12}$.

    d. There are 36 outcomes in the sample space and three ways to get a sum of 3 or 2: (1, 2), (2, 1), and (1, 1). Hence, $P(\text{sum less than 4}) = \dfrac{3}{36} = \dfrac{1}{12}$.

3. a. There are 52 outcomes in the sample space and one 7 of spades; hence, $P(\text{seven of spades}) = \dfrac{1}{52}$.

    b. There are 52 outcomes in the sample space and four jacks; hence, $P(\text{jack}) = \dfrac{4}{52} = \dfrac{1}{13}$.

    c. There are 52 outcomes in the sample space and 13 clubs; hence, $P(\text{club}) = \dfrac{13}{52} = \dfrac{1}{4}$.

    d. There are 52 outcomes in the sample space and 13 hearts and 13 clubs; hence, $P(\text{heart or club}) = \dfrac{13 + 13}{52} = \dfrac{26}{52} = \dfrac{1}{2}$.

    e. There are 52 outcomes in the sample space and 26 red cards (13 diamonds and 13 hearts); hence, $P(\text{red card}) = \dfrac{26}{52} = \dfrac{1}{2}$.

4. The sample space for three children is

   BBB   GGB

   BBG   GBG

   BGB   BGG

   GBB   GGG

   a. There are eight outcomes in the sample space and one way to get all girls: GGG; hence, $P(\text{all girls}) = \dfrac{1}{8}$.

   b. There are eight outcomes in the sample space and two ways to get all boys or all girls: BBB and GGG; hence, $P(\text{all boys or all girls}) = \dfrac{2}{8} = \dfrac{1}{4}$.

   c. There are eight outcomes in the sample space and three ways to get two girls and one boy: GGB, GBG, BGG; hence, $P(\text{exactly 2 girls and 1 boy}) = \dfrac{3}{8}$.

5. There are 36 outcomes in the sample space and five ways to get an 8 and three ways to get a 10; hence, $P(8 \text{ or } 10) = \dfrac{5+3}{36} = \dfrac{8}{36} = \dfrac{2}{9}$.

6. There are $9 + 6 = 15$ cans in the cooler and 9 of them are cola; hence $P(\text{cola}) = \dfrac{9}{15} = \dfrac{3}{5}$.

7. There are 36 outcomes in the sample space and 10 ways to get a sum greater than 8. There are 6 ways to get doubles, but $(5, 5)$ and $(6, 6)$ have been counted twice, so

   $P(\text{sum greater than 8 or doubles}) = \dfrac{10}{36} + \dfrac{6}{36} - \dfrac{2}{36} = \dfrac{14}{36} = \dfrac{7}{18}$.

8. $P(2 \text{ orange balls}) = P(\text{orange}) \times P(\text{orange, given that an orange ball has occurred}) = \dfrac{3}{6} \times \dfrac{2}{5} = \dfrac{\cancel{3}^{1}}{\cancel{6}_{\cancel{3}^{1}}} \times \dfrac{\cancel{2}^{1}}{5} = \dfrac{1}{5}$.

9. $P(3 \text{ even numbers}) = \dfrac{1}{2} \times \dfrac{1}{2} \times \dfrac{1}{2} = \dfrac{1}{8}$. The events are independent.

10. In this case, any number can occur the first time, but on the second roll, the outcome has to match the number that occurred the first time. That is $\dfrac{1}{6}$. Hence, $P(\text{same number twice}) = \dfrac{\cancel{6}^{1}}{\cancel{6}^{1}} \times \dfrac{1}{6} = \dfrac{1}{6}$.

In this section, you learned to solve simple probability problems. Here the solutions are obtained by determining the number of outcomes in the sample space. This number is placed in the denominator of the fraction. The number of outcomes desired is placed in the numerator of the fraction. The fraction is reduced if possible. Several probability rules were presented in this section and rules for determining the probability of events when *or* is used are given. Finally, when the probability experiment is performed more than once, two additional rules were explained.

Solving Statistics Problems

Statistics is the science of conducting studies to collect, organize, analyze, summarize, and draw conclusions from data. The *data* can be numbers such as weights, temperatures, test scores, etc., or observations such as colors of automobiles, political affiliations, etc. A group of data values collected for a particular study is called a *data set*. Statistics is used in almost all fields of human endeavor.

In statistics, there are three commonly used measures of average. They are the mean, median, and mode.

The *mean* is the sum of the data values divided by the total number of data values.

EXAMPLE

Find the mean of 9, 23, 15, 20, and 18.

SOLUTION

Goal: You are being asked to find the mean for the given data set.

Strategy: Add the values and divide the sum by 5 (there are five data values).

Implementation:

$$9 + 23 + 15 + 20 + 18 = 85$$
$$85 \div 5 = 17$$

The mean is 17.

Evaluation: The mean will fall between the lowest and highest values and, most of the time, somewhere near the middle of the values.

The *median* is a value that falls in the center of the data set. You must first arrange the data in order from the smallest data value to the largest data value.

EXAMPLE

Find the median for 17, 24, 22, 16, and 7.

SOLUTION

Goal: You are being asked to find the median for the given data set.

Strategy: Arrange the data values in order and find the middle value.

Implementation:

$$7, 16, 17, 22, 24$$

Since 17 is the middle value, the median is 17.

Evaluation: Check to see if the data values are arranged correctly; then make sure you have found the middle value.

If the number of data values is odd, as in the previous example, the median will be one of the values; however, if the number of data values is even, the median will fall halfway between the middle two values, as shown in the next example.

EXAMPLE

Find the median for 86, 23, 52, 63, 44, and 91.

SOLUTION

Goal: You are being asked to find the median for the given data set.

Strategy: Arrange the data in order; then find the middle point.

Implementation:

$$23, 44, 52, 63, 86, 91$$

The middle of the data is halfway between 52 and 63; hence, the median is

$$\frac{52 + 63}{2} = \frac{115}{2} = 57.5$$

Evaluation: Check the solution.

The third measure of average is called the mode. The *mode* is the data value that occurs most often.

 EXAMPLE

Find the mode of 19, 24, 16, 18, 19, and 27.

 SOLUTION

Goal: You are being asked to find the mode for the given data set.

Strategy: Find the value that occurs most often.

Implementation:

It is helpful, although not necessary, to arrange the data in order:

16, 18, 19, 19, 24, 27

Since 19 occurs twice and that is more often than any other number, 19 is the mode.

Evaluation: The answer is obvious.

EXAMPLE

Find the mode for 5, 6, 8, 9, 9, 9, 10, 10, 12, 12, 12, and 16.

SOLUTION

Goal: You are being asked to find the mode for the given data set.

Strategy: Analyze the data and see what value occurs most often.

Implementation: In this case, the values of 9 and 12 occur three times. Hence, the data has two modes. They are 9 and 12.

Evaluation: The answer is obvious.

 EXAMPLE

Find the mode for 103, 206, 87, 54, and 153.

SOLUTION

Goal: You are being asked to find the mode for the given data set.

Strategy: Find the data value that occurs most often.

Implementation: In this case, each data value occurs only once. Hence, we say that there is no mode.

Two things should be noted:

1. The mode of a data set can be a single value, more than one value, or no value at all.

2. The mean, median, and mode for a data set, in most cases, will not be equal.

In addition to the measures of average, statisticians also use measures of variation to describe a data set. The two most often used measures of variation are the range and the standard deviation. These measures describe the spread of the data about the mean. Loosely speaking, the larger the range or standard deviation, the more variable or spread out the data is in the set.

The *range* is found by subtracting the smallest data value from the largest data value.

EXAMPLE

Find the range for 17, 32, 19, 16, and 15.

SOLUTION

Goal: You are being asked to find the range for the given data set.

Strategy: Subtract the smallest data value from the largest data value in the set.

Implementation: The smallest data value is 15, and the largest data value is 32, so the range is 32 – 15 = 17.

Evaluation: Redo the problem.

The range is a rough estimate of variation, so statisticians also use what is called the *standard deviation*. The standard deviation can be computed by using the following procedure:

1. Find the mean for the data set.
2. Subtract the mean from each value in the data set.
3. Square the differences.
4. Find the sum of the squares.
5. Divide the sum by $n - 1$, where n is the number of data values.
6. Take the square root of the answer. (You may need a calculator for this step.)

EXAMPLE

Find the standard deviation: 14, 22, 16, 28, and 20.

SOLUTION

Goal: You are being asked to find the standard deviation for the given data set.

Strategy: Use the procedure given previously.

Implementation:

1. Find the mean:

    $$14 + 22 + 16 + 28 + 20 = 100$$

    $$100 \div 5 = 20$$

2. Subtract the mean from each data value:

    $$14 - 20 = -6$$

    $$22 - 20 = 2$$

    $$16 - 20 = -4$$

    $$28 - 20 = 8$$

    $$20 - 20 = 0$$

3. Square the answers:

    $$(-6)^2 = 36$$

    $$2^2 = 4$$

    $$(-4)^2 = 16$$

    $$8^2 = 64$$

    $$0^2 = 0$$

4. Find the sum of the squares:

    $$36 + 4 + 16 + 64 + 0 = 120$$

5. Divide the sum by $n - 1$, where $n = 5$ and $n - 1 = 5 - 1 = 4$:

    $$120 \div 4 = 30$$

6. Find the square root of 30:

    $$\sqrt{30} = 5.48 \text{ (rounded)}$$

    The standard deviation is 5.48.

Evaluation: The standard deviation can be estimated by dividing the range by 4. In this case, the range is $28 - 14 = 14$. Thus, $14 \div 4 = 3.5$. Since this is only a rough estimate, we are in the ball park.

Roughly speaking, most of the data values will usually fall between two standard deviations of the mean.

TRY THESE

For the data set 28, 13, 19, 24, 18, and 24, find each:

1. The mean

2. The median

3. The mode

4. The range

5. The standard deviation

SOLUTIONS

1. $28 + 13 + 19 + 24 + 18 + 24 = 126$ $126 \div 6 = 21$

   The mean = 21.

2. 13, 18, 19, 24, 24, 28

   The middle value is halfway between 19 and 24; hence, the median is $(19 + 24) \div 2 = 43 \div 2 = 21.5$.

3. The value that occurs most often is 24, so the mode is 24.

4. The range is $28 - 13 = 15$.

5. To find the standard deviation, follow these steps:

   Find the mean. It is 21, as found in answer 1.

   Subtract the mean from each data value:

   $28 - 21 = 7$

   $13 - 21 = -8$

   $19 - 21 = -2$

   $24 - 21 = 3$

   $18 - 21 = -3$

   $24 - 21 = 3$

   Square the differences:

   $7^2 = 49$

   $(-8)^2 = 64$

   $(-2)^2 = 4$

   $3^2 = 9$

   $(-3)^2 = 9$

   $3^2 = 9$

Find the sum of the differences:

$49 + 64 + 4 + 9 + 9 + 9 = 144$

Divide the sum by $6 - 1 + 5$

$$\frac{144}{5} = 28.8$$

Find the square root of 28.8:

$\sqrt{28.8} = 5.37$ (rounded)

Hence, the standard deviation is 5.37.

In this section, you learned how to solve problems using statistics. There are three measures of average. They are the mean, median, and mode. There are two measures of variation. They are the range and standard deviation. These are the common statistical measures that are most often used.

Summary

This chapter explained how to solve three special types of problems. They are geometry problems, probability problems, and statistics problems. Geometry problems use basic geometric principles. Probability and statistics problems use formulas.

QUIZ

1. A single card is selected from a deck of cards. Find the probability that it is a club.

   A. $\dfrac{1}{4}$

   B. $\dfrac{1}{13}$

   C. $\dfrac{1}{2}$

   D. $\dfrac{1}{8}$

2. Two dice are rolled; find the probability of getting a sum of 11 or a sum of less than 4.

   A. $\dfrac{1}{6}$

   B. $\dfrac{1}{12}$

   C. $\dfrac{5}{36}$

   D. $\dfrac{1}{8}$

3. A single die is rolled; find the probability of getting a 7.

   A. 0

   B. $\dfrac{1}{6}$

   C. 1

   D. Cannot be computed

4. Find the mean of 156, 170, 192, and 146.

   A. 166

   B. 142

   C. 163

   D. 175

5. Find the median of 12, 5, 10, and 16.

   A. 10

   B. 7.5

   C. 10.5

   D. 11

6. Find the median of 56, 18, 44, 22, and 65.
   A. 41.5
   B. 44
   C. 40
   D. 48.2

7. Find the mode of 19, 37, 15, 14, and 18.
   A. 19
   B. 18
   C. 20.6
   D. no mode

8. Find the mode of 6, 5, 8, 4, 5, 9, and 12.
   A. 5
   B. 8.5
   C. 8
   D. 7

9. Find the range of 8, 14, 10, 8, and 22.
   A. 10
   B. 14
   C. 8
   D. 22

10. Find the standard deviation (rounded to one decimal place) of 34, 36, 24, 18, 26.
    A. 3.6
    B. 5.2
    C. 13.7
    D. 7.4

In New Jersey government owns 129,79... cres of land. In Texas, ...ment own... 2,307,171 acres ... land and in ... la... go...rnment o... ...213... of land. Find the total amount of land own... ...y the federal go...ernment in the th...ee states.

Final Exam

1. The size of Cuba is 42,031 square miles, and the size of Great Britain is 88,407 square miles. How much larger is Great Britain?

   A. 46,376 square miles

   B. 54,327 square miles

   C. 35,162 square miles

   D. 130,438 square miles

2. Find the total of the areas of the Sea of Japan, which is 391,100 square miles, and the Hudson Bay, which is 281,900 square miles.

   A. 109,200 square miles

   B. 673,000 square miles

   C. 432,100 square miles

   D. 323,300 square miles

3. If a person pays $324 a month on a loan, how much will the person pay in a year?

   A. $27

   B. $336

   C. $3,888

   D. $5,428

4. How many boxes are needed to package 448 bottles of shampoo if 14 bottles can fit in a box?

   A. 32

   B. 16

   C. 28

   D. 6,272

5. A person traveled from her home to a bakery, a distance of $8\frac{5}{8}$ miles. Then she went to her salon, a distance of $4\frac{1}{3}$ miles from the bakery. How far did she travel in all?

   A. $15\frac{7}{8}$ miles

   B. $12\frac{23}{24}$ miles

   C. $6\frac{3}{8}$ miles

   D. $4\frac{7}{24}$ miles

6. A generator uses $\frac{4}{5}$ gallon of gasoline per hour. How many gallons of gasoline are used if it is run $3\frac{3}{4}$ hours?

   A. $4\frac{11}{20}$ gallons

   B. 3 gallons

   C. $2\frac{19}{20}$ gallons

   D. 5 gallons

7. A taxi service charges $8 plus 75 cents per mile to rent a taxi. How much does a person pay for a 16-mile trip?

   A. $12

   B. $8.75

   C. $14.25

   D. $20

8. How many pieces of wood $1\frac{2}{3}$ feet long can be cut from a board that is 10 feet long?

   A. 3

   B. 4

   C. 5

   D. 6

9. A clerk sold $3\frac{1}{4}$ pounds of peanuts, $5\frac{3}{8}$ pounds of cashews, and $2\frac{1}{2}$ pounds of almonds. How many pounds of nuts were sold in all?

   A. $10\frac{2}{3}$ pounds

   B. $9\frac{5}{6}$ pounds

   C. $11\frac{1}{8}$ pounds

   D. $8\frac{1}{8}$ pounds

10. Mike is $5\frac{3}{4}$ feet tall and Cindy is $4\frac{7}{8}$ feet tall. How much taller is Mike?

    A. $\frac{7}{8}$ feet

    B. $1\frac{5}{8}$ feet

    C. $10\frac{5}{8}$ feet

    D. $1\frac{1}{8}$ feet

11. A person traveled 374.4 miles on 16 gallons of gasoline. How many miles per gallon did the person get?

    A. 22.6 miles per gallon

    B. 23.4 miles per gallon

    C. 21.5 miles per gallon

    D. 24.7 miles per gallon

12. Julie's bicycle speedometer read 534.2 miles before she started her ride. When she finished, her speedometer read 551.6 miles. How far did she travel?

   A. 18.4 miles

   B. 14.4 miles

   C. 17.4 miles

   D. 16.4 miles

13. The value of a home has increased 15%. How much is the home worth now if its original price was $71,875?

   A. $10,781.25

   B. $68,475

   C. $62,500

   D. $82,656.25

14. A person receives a 4% raise. Find the new salary if he earns $32,000 now.

   A. $1,280

   B. $28,250

   C. $33,280

   D. $30,100

15. A person drove 386.1 miles and got 23.4 miles per gallon. How many gallons did the person use?

   A. 16.5

   B. 18.3

   C. 23.4

   D. 20.5

16. What is the selling price of a camera if the sales tax is $14.52 and the rate is 6%?

   A. $276.98

   B. $242

   C. $87.12

   D. $321

17. In order to get a light blue paint, 2 gallons of white paint are mixed with 5 gallons of blue paint. To get the same color, how many gallons of white paint are needed to be mixed with 22 gallons of blue paint?

    A. 8.8 gallons

    B. 6.4 gallons

    C. 7 gallons

    D. 6 gallons

18. Mike bought 14 candy bars and paid $25.50. If some of the bars cost $1.25 and the rest cost $2.25, how many of the $2.25 candy bars did he buy?

    A. 6

    B. 4

    C. 5

    D. 8

19. On a map, the scale is $\frac{3}{4}$ inch = 30 miles. Find the actual distance between two cities if they are 3 inches apart.

    A. 60 miles

    B. 54.8 miles

    C. 72 miles

    D. 120 miles

20. Three years ago, Harry was twice as old as his brother. If the difference in their ages is 8 years, how old is Harry today?

    A. 10

    B. 15

    C. 19

    D. 21

21. One pipe can fill a tank in 16 hours and another pipe can fill the tank in 20 hours. If both pipes are opened, how long will it take to fill the tank? (Round the answer to one decimal place.)

    A. 12.3 hours

    B. 10.6 hours

    C. 9.3 hours

    D. 8.9 hours

22. A lever is 10 feet long. Where should the fulcrum be placed in order to balance 40 pounds at one end and 160 pounds from the other end?

    A. 6 feet from the 40 pounds

    B. 8 feet from the 40 pounds

    C. 3 feet from the 40 pounds

    D. 5 feet from the 40 pounds

23. If the product of two positive consecutive even numbers is 3,024, find the larger number.

    A. 46

    B. 54

    C. 56

    D. 48

24. If the length of a rectangular platform is 6 feet more than its width and the area of the platform is 391 square feet, find the length of the platform.

    A. 15 feet

    B. 23 feet

    C. 19 feet

    D. 17 feet

25. Lori has two savings accounts. One account pays 4.6% interest and the other pays 2.5%. If the total investment is $19,000 and the total interest is $664, find the amount of money Lori has invested at 4.6%.

    A. $8,000

    B. $9,000

    C. $5,200

    D. $6,500

26. A child's bank contains 42 coins consisting of nickels and quarters only. Find the number of nickels it contains if the total amount in the bank is $5.10.

    A. 15

    B. 19

    C. 27

    D. 31

27. The sum of the digits of a two-digit number is 12. If the digits are reversed, the new number is 36 more than the original number. Find the number.

    A. 48

    B. 39

    C. 93

    D. 84

28. In a two-digit number, the tens digit is 5 more than the ones digit. If the digits are reversed, the new number is 45 less than the original number. Find the original number.

    A. 72

    B. 61

    C. 94

    D. 83

29. An airplane took 10 hours to fly a distance of 750 miles, flying against the wind. If the return trip took 6 hours flying with the wind, find the speed of the wind.

    A. 20 miles per hour

    B. 25 miles per hour

    C. 15 miles per hour

    D. 30 miles per hour

30. Two people leave two towns that are 200 miles apart and drive toward each other. If one person drives 8 miles per hour slower than the other, and they meet in two hours, how fast was the slower driver going?

    A. 42 miles per hour

    B. 46 miles per hour

    C. 50 miles per hour

    D. 54 miles per hour

31. How far will an automobile travel in $3\frac{3}{8}$ hours at a speed of 32 miles per hour? (Use $D = RT$.)

    A. 96 miles

    B. 99 miles

    C. 108 miles

    D. 116 miles

32. Find the interest on a loan of $9650 at 4% for seven years. (Use $I = PRT$.)

   A. $386

   B. $2,702

   C. $3,160

   D. $4,825

33. Find the area of a triangle whose base is 16 feet and whose height is 9 feet. (Use $A = \dfrac{1}{2}bh$.)

   A. 72 square feet

   B. 144 square feet

   C. 25 square feet

   D. 50 square feet

34. Find the distance an object falls in 12 seconds. (Use $d = \dfrac{1}{2}(32)t^2$.)

   A. 4,608 feet

   B. 192 feet

   C. 96 feet

   D. 2,304 feet

35. Two angles of a triangle are equal in measure. If the third angle is 15° greater than the other angles, find the measure of the third angle. The sum of the measures of the angles of a triangle is 180°.

   A. 70°

   B. 55°

   C. 50°

   D. 65°

36. A store owner has eight more scarves than she has jackets. Find the number of jackets she has if she has a total of 40 items.

   A. 8

   B. 16

   C. 20

   D. 24

37. When two dice are rolled, the probability of getting a sum of 10 is

    A. $\dfrac{1}{36}$

    B. $\dfrac{1}{8}$

    C. $\dfrac{1}{12}$

    D. $\dfrac{1}{9}$

38. The probability of getting an 8 when a single die is rolled is

    A. $\dfrac{1}{6}$

    B. 0

    C. $\dfrac{8}{6}$

    D. 1

39. When a card is selected from a deck, the probability of getting an ace and a red card is

    A. $\dfrac{1}{26}$

    B. $\dfrac{1}{13}$

    C. $\dfrac{1}{4}$

    D. $\dfrac{1}{52}$

40. A committee consists of five women and four men. If a chairperson is selected, find the probability that it is a woman.

    A. $\dfrac{1}{5}$

    B. $\dfrac{1}{4}$

    C. $\dfrac{4}{9}$

    D. $\dfrac{5}{9}$

41. When three coins are tossed, the probability of getting 0, 1, 2, or 3 heads is

    A. 0

    B. $\dfrac{1}{2}$

    C. $\dfrac{3}{4}$

    D. 1

42. The sum of the probabilities of all the events in the sample space will always be

    A. 0

    B. 1

    C. $\dfrac{1}{2}$

    D. It varies

43. A professor has 10 books on a shelf in his office. Three are calculus books, two are algebra books, and five are statistics books. If he selects a book at random, what is the probability that he selects an algebra book or a statistics book?

    A. $\dfrac{7}{10}$

    B. $\dfrac{1}{2}$

    C. $\dfrac{1}{5}$

    D. $\dfrac{3}{10}$

44. Find the mean of 19, 27, 14, 19, 16, 20, and 18.

    A. 18

    B. 18.5

    C. 19

    D. No mean

45. Which of the following statements is true?

    A. The mean, median, and mode of a data set will always be equal.

    B. The mean, median, and mode of a data set can never be equal.

C. If the data in the data set are whole numbers, the mean will always be a whole number.

D. None of the above statements is true.

46. **Find the median of 42, 87, 16, 23, 27, 52, 63, and 20.**

   A. 41.25

   B. 42

   C. 27

   D. 34.5

47. **Find the mode of 20, 7, 19, 11, 17, and 19.**

   A. 17.5

   B. 15

   C. 19

   D. 16.5

48. **Find the mode of 32, 52, 43, 38, and 41.**

   A. No mode

   B. 43

   C. 42

   D. 38

49. **Find the range of 38, 52, 75, 19, 63, and 37.**

   A. 31

   B. 56

   C. 14

   D. 49

50. **Find the standard deviation rounded to two places of 12, 18, 20, 23, and 27.**

   A. 20

   B. 5.61

   C. 31.5

   D. 19.5

Answers to Quizzes and Final Exam

Chapter 1	Chapter 3	Chapter 5	Chapter 7	Chapter 9	Chapter 11
1. B	1. C	1. C	1. B	1. B	1. A
2. D	2. C	2. D	2. D	2. B	2. D
3. D	3. C	3. A	3. C	3. D	3. A
4. D	4. C	4. B	4. B	4. C	4. B
5. C	5. B	5. C	5. A	5. D	5. C
6. A	6. D	6. C	6. B	6. A	6. D
7. B	7. A	7. B	7. C	7. C	7. A
8. C	8. B	8. A	8. B	8. B	8. B
9. D	9. D	9. D	9. D	9. C	9. C
10. C	10. A	10. B	10. D	10. D	10. C

Chapter 2	Chapter 4	Chapter 6	Chapter 8	Chapter 10	Chapter 12
1. B	1. C	1. B	1. A	1. C	1. A
2. A	2. D	2. A	2. D	2. B	2. C
3. C	3. A	3. C	3. B	3. B	3. A
4. D	4. B	4. D	4. C	4. D	4. A
5. A	5. D	5. C	5. B	5. A	5. D
6. D	6. C	6. B	6. A	6. C	6. B
7. B	7. B	7. A	7. C	7. D	7. D
8. C	8. C	8. D	8. D	8. A	8. A
9. D	9. A	9. B	9. C	9. B	9. B
10. B	10. D	10. A	10. C	10. C	10. D

Final Exam

1. A	11. B	21. D	31. C	41. D
2. B	12. C	22. B	32. B	42. B
3. C	13. D	23. C	33. A	43. A
4. A	14. C	24. B	34. D	44. C
5. B	15. A	25. B	35. A	45. D
6. B	16. B	26. C	36. B	46. D
7. D	17. A	27. A	37. C	47. C
8. D	18. D	28. D	38. B	48. A
9. C	19. D	29. B	39. A	49. B
10. A	20. C	30. B	40. D	50. B

Suggestions for Success in Mathematics

1. Be sure to attend every class. If you know ahead of time that you will be absent, tell your instructor and get the assignment. If it is an emergency absence, get the assignment from another student. Try to do the problems before the next class. If possible, get the class notes from another student.

2. Read the material in the textbook several times. Write down or underline all definitions, rules, and symbols. Try to do the sample problems.

3. Do all assigned homework as soon as possible before the next class. Concentrate on mathematics only. Get all of your materials before you start doing your homework. Make sure you write the assignment on the top of your homework. Read the directions. Copy each problem on your homework paper. Make sure that you have copied it correctly. Do *not* use scratch paper. Work out each problem in detail and do not skip steps. Write neatly and large enough. Check the answer with the one in the back of the book or rework the problem again. If you did not get the correct answer, try to find your mistake or start over. Don't look for shortcuts, because they do not always work. Write down any questions you have and ask your instructor or another student at the next class period. If you are having difficulty with the problem, consult your textbook and notes. Don't give up too quickly.

4. Always review before each exam. You can usually find a review or chapter test at the end of each chapter in the book. If not, you can make up your

own review by selecting several problems from each section in the book to try. If you can't get the correct answer, ask the teacher or another student to help you before the exam. If you have made study cards, review them.

5. On the day of the test, arrive early. Look over your notes and study cards. Bring all necessary materials such as pencils, protractor, calculator, textbook, etc., to class. When you get the test, look over the entire test before you get started. Read the directions. Work the problems that you know how to do first. Do not spend too much time on any one problem. After you have finished the test, if time allows, check each problem. When you get the test back, check your mistakes and study the types of problems that you have missed, because similar problems may be on the final exam.

6. If you have difficulty with mathematics, arrange for a tutor. Some schools have learning centers where you can receive free tutoring.

7. Finally, make sure that you are in the correct class. You cannot skip math classes. Mathematics is sequential in nature. What you learn today, you will use tomorrow. What you learn in one course, you will use in the next course.

GOOD LUCK!

Index